Springer Optimization and Its Applications

Volume 215

Aims and Scope

Optimization has continued to expand in all directions at an astonishing rate. New algorithmic and theoretical techniques are continually developing and the diffusion into other disciplines is proceeding at a rapid pace, with a spot light on machine learning, artificial intelligence, and quantum computing. Our knowledge of all aspects of the field has grown even more profound. At the same time, one of the most striking trends in optimization is the constantly increasing emphasis on the interdisciplinary nature of the field. Optimization has been a basic tool in areas not limited to applied mathematics, engineering, medicine, economics, computer science, operations research, and other sciences.

The series **Springer Optimization and Its Applications (SOIA)** aims to publish state-of-the-art expository works (monographs, contributed volumes, textbooks, handbooks) that focus on theory, methods, and applications of optimization. Topics covered include, but are not limited to, nonlinear optimization, combinatorial optimization, continuous optimization, stochastic optimization, Bayesian optimization, optimal control, discrete optimization, multi-objective optimization, and more. New to the series portfolio include Works at the intersection of optimization and machine learning, artificial intelligence, and quantum computing.

Volumes from this series are indexed by Web of Science, zbMATH, Mathematical Reviews, and SCOPUS.

Manjunath Patel G. C. • Ajith B. S.
Jagadish • Arun Kumar Shettigar
Olusegun David Samuel
Editors

Biofuel Production, Performance, and Emission Optimization

A Comprehensive Approach to Modelling and Optimization

Springer

Manjunath Patel G. C.
Department of Mechanical Engineering
PES Institute of Technology and
Management, Shivamogga, Visvesvaraya
Technological University
Belagavi, Karnataka, India

Ajith B. S.
Department of Mechanical Engineering
Sahyadri College of Engineering &
Management, Mangaluru, Visvesvaraya
Technological University
Belagavi, Karnataka, India

Jagadish
Statistical Quality Control & Operations
Research Unit
Indian Statistical Institute
Bangalore, Karnataka, India

Arun Kumar Shettigar
Department of Mechanical Engineering
National Institute of Technology Karnataka
Surathkal, Karnataka, India

Olusegun David Samuel
Department of Mechanical Engineering
Fed. University of Petroleum Resources
Effurun, Nigeria

ISSN 1931-6828 ISSN 1931-6836 (electronic)
Springer Optimization and Its Applications
ISBN 978-3-031-84805-6 ISBN 978-3-031-84806-3 (eBook)
https://doi.org/10.1007/978-3-031-84806-3

This Springer imprint is published by the registered company Springer Nature Switzerland AG
The registered company address is: Gewerbestrasse 11, 6330 Cham, Switzerland

This book is dedicated to all my teachers and friends

Foreword

This book proposes the realms of biodiesel research introduced to the real world, focusing on a systematic framework for experimentation, modeling, and optimization. This book highlights the following:

- Importance of biofuel to meet the stringent global energy demand.
- Salient points to consider for selecting feedstock for biodiesel production.
- Selection criteria in evaluating the methods of oil extraction and biodiesel conversion.
- Experimental methods in analyzing and optimizing biodiesel production.
- The cost incurred for biodiesel conversion from feedstock is estimated.
- Proposed a framework for experimentation, modeling, analysis, and optimization.
- Measurement procedure with ASTM standards and results of various characterization of oils and biodiesel.
- Optimized transesterification condition on the performance of biodiesel conversion with and without Ag_2O nanoparticles.
- Overall uncertainty in experiments was estimated.
- Experimental analysis and optimization of engine parameters for better performance and emission characteristics.
- Novel empirical equations were derived for all performance and emission characteristics using a central composite design of experiments.
- Artificial neural networks modeling of engine performance and emission characteristics.
- The importance of optimizing neural network parameters in reducing training error is discussed.
- The reverse modeling task, which involves predicting the input parameters for desired performance and emission characteristics in diesel engines, is discussed in detail due to its significant role in the research.
- Mathematical formulation for conducting multiple objective optimizations is proposed by highlighting the importance of assigning weight fractions.

- Six meta-heuristic algorithms and their importance in tuning algorithm-specific and common parameters were discussed.
- The input variables were optimized simultaneously for conflicting performance and emission characteristics requirements.

The above information inspires readers with detailed insights into process kinetics and mechanics of the transesterification process for biodiesel conversion and engine parameters for emission reduction and improving diesel engine performances without significant modification. The optimized conditions could help readers enhance biodiesel production and engine performance and reduce emissions.

Karnataka, India Manjunath Patel G. C.

Preface

Fossil fuels-derived non-renewable energy sources satisfy 84% of the energy demand, and they're extinct at an unprecedented rate, pollution problems led to a search for alternate sources. This book presents the details of potential renewable energy sources that satisfy the energy demand by addressing the potential challenges in the selection of sustainable feedstocks (~70–95% of biodiesel price is accounted for feedstock alone) and process (oil extraction and biofuel conversion) associated to technologies, investment and infrastructure suitable for industrial-scale production. To ensure sustainability, the selected feedstock must be abundantly available, economy, capable of yielding high oil, ease of processing, higher biofuel conversion, better fuel properties close to diesel fuel, and by-products must possess higher commercial value. Garcinia gummi-gutta seeds are identified as a sustainable feedstock for biodiesel production. Regarding biodiesel conversion from seed oil, the selection and optimization of transesterification variables for higher yield are of industrial relevance.

Furthermore, the practical usefulness of biofuel properties is tested to run in diesel engines without modification. The conflicting requirements in engine performance and emission characteristics (i.e. maximize: brake thermal efficiency, minimize: brake-specific fuel consumption, carbon monoxide emission, unburnt hydrocarbon, and nitrous oxide emission) require optimization of parameters for sustainable benefits. Research, development, and innovations in the field lead to adopting new-age tools and techniques (statistical design of experiments: for efficient data collection and artificial intelligence and machine learning algorithms for prediction and optimization) for detailed insight analysis and understanding of the process's physics kinetics and mechanics. This book provides insights into feedstock selection, oil extraction, biodiesel conversion, and evaluation of engine performance and emission characteristics fuelled with biodiesel and its blend. This includes the outcome of the research carried out by authors in the form of experimental results, analysis, statistical and intelligent modelling, and optimization. The proposed book aims to help novel and professional engineers by facilitating the framework from feedstock selection to optimizing engine performance and emission characteristics. Chapter 1 begins with an introduction to renewable energy resources and sustainable feedstocks for biodiesel. Chapter 2 discusses the previous studies on experimentation, modelling and optimization of oil extraction, biodiesel production methods, and engine performance and emission characteristics. Results of an experimental investigation on biodiesel production, fuel properties, engine performance, and emission characteristics conducted by authors are reported, analyzed, and discussed in Chap. 3. The chapter also addresses factors' effect on biodiesel yield, engine performances, and emission characteristics. Chapter 4 is focused on artificial neural network-based intelligent modelling to establish the relationship between input and output parameters. This chapter also discusses the prediction of input parameters through reverse modelling to get the desired emission and performance characteristics. Chapter 5 discusses the optimization for the higher biodiesel conversion process and desired performance and emission characteristics using artificial intelligence tools.

The results of experimental investigation, new-age technological tools employed for biodiesel production and their performance and emission characteristics conducted by authors are reported in this book. The authors hope that the systematic framework written for experimentation, modelling, and optimization would facilitate and motivate novel or professional researchers/engineers in biodiesel production at an industrial scale and perform engine optimization.

Shivamogga, Karnataka, India — Manjunath Patel G. C.
Mangaluru, Karnataka, India — Ajith B. S.
Bangalore, Karnataka, India — Jagadish
Surathkal, Karnataka, India — Arun Kumar Shettigar
Effurun, Nigeria — Olusegun David Samuel
Shivamogga Karnataka

Contents

List of Abbreviations

A:F	Air-to-fuel ratio
ABC	Artificial bee colony
AC	Ant colony optimization
Ag_2O	Silver oxide
a_H	Activation function constant for hidden layer
AI	Artificial intelligence
Al_2O_3	Aluminium oxide
ANN	Artificial neural network
ANOVA	Analysis of variance
a_O	Activation function constant for output layer
ASTM	American Society for Testing and Materials
b	Bias
B10	10% volume of biodiesel + 90% volume of diesel
B100	100% volume of biodiesel
B20	20% volume of biodiesel + 80% volume of diesel
B30	30% volume of biodiesel + 70% volume of diesel
BB-BC	Big bang big crunch
BBD	Box-Behnken design
BFO	Bacteria foraging optimization
BP	Brake power
BPNN	Back-propagation neural network
BSFC	Brake-specific fuel consumption
$BSFC_{Max}$	Maximum value of BSFC
$BSFC_{Min}$	Minimum value of BSFC
BT	Brake torque
BTE	Brake thermal efficiency
BTE_{Max}	Maximum value of BTE
BTE_{Min}	Minimum value of BTE
CaO	Calcium oxide
CBD	Compression before diesel
CC	Catalyst concentration

CCC	Central composite circumscribe
CCD	Central composite design
CCF	Central composite face centred
CCI	Central composite inscribe
CE	Standard error coefficient
CEEA	Classical engineering experimental approach
CH_3OK	Potassium methoxide
CH_3ONa	Sodium methoxide
CN	Cetane number
CNMNT	Cetane number of the 1-methylnaphthalene
CO	Carbon monoxide
CO_{Max}	Maximum value of CO
CO_{Min}	Minimum value of CO
C-o-R	Coefficient of restitution
CR	Compression ratio
D100	100% volume of diesel
DE	Differential evolution
DF	Degrees of freedom
DO	Overall desirability
DOE	Design of experiments
DZCD	Delayed zone combustion duration
DZHRC	Delayed hydrocarbon reduction control
EBOA	Election-based optimization algorithm
EDS	Energy-dispersive spectroscopy
EGT	Exhaust gas temperature
EL	Engine load
EM	Experimental method
ES	Engine speed
FA	Firefly Algorithm
FAME	Fatty acid methyl ester
FFD	Full factorial design
FFR	Fluid flow rate
FTIR	Fourier transform infrared spectroscopy
FWHM	Full width at half maximum
GA	Genetic algorithm
GA-NN	Genetic algorithm–neural network
GC-MS	Gas chromatography-mass spectrometry
GGG	Garcinia gummi-gutta
GOA	Grasshopper optimization algorithm
GRA	Grey relational analysis
GWO	Grey wolf optimizer
H_2SO_4	Sulphuric acid
HC	Hydrocarbon
HCl	Hydrochloric acid
HS	Harmony search

IA	Injection advance
ID	Iodine value
IR spectroscopy	Infrared spectroscopy
IT	Ignition timing
KOH	Potassium hydroxide
M:O	Methanol-to-oil molar ratio
MAE	Microwave-assisted extraction
MAPE	Mean absolute percentage error
MgO	Magnesium oxide
ML	Machine learning
MLP	Multilayer perceptron
MOO	Multi-objective optimization
MSE	Mean squared error
Ni-ZnO	Nickel-doped zinc oxide
NOx	Nitrogen oxide
NOx_{Max}	Maximum value of NOx
NOx_{Min}	Minimum value of NOx
NPs	Nanoparticles
NSGA-II	Non-dominated Sorting Genetic Algorithm II
O_2	Oxygen
OF	Objective function
OFAT	One-Factor-At-A-Time
OM	Optimization method
OU	Overall uncertainty
PD	Percent deviation
PM	Particulate matter
PR	Power ratio
PSO	Particle swarm optimization
PT	Power torque
PZCD	Peak zone combustion duration
PZHRC	Peak hydrocarbon reduction control
$R_1,...,R_6$	Response 1 to 6
R^2	Coefficient of determination
RCCD	Rotatable Central Composite Design
RES	Renewable energy sources
RPM	Revolution per minute
RSM	Response surface methodology
SE Coef	Standard error coefficient
SEM	Scanning electron microscopy
SFE	Supercritical fluid extraction
SL	Stroke length
SnO_2	Tin dioxide
SO_2	Sulphur dioxide
SrO	Strontium oxide
TiO_2	Titanium dioxide

TLBO	Teaching-learning-based optimization
UHC	Unburnt hydrocarbon
UHC_{Max}	Maximum value of UHC
UHC_{Min}	Minimum value of UHC
$V_1,...,V_4$	Variable 1 to Variable 4
$W_1,...,W_5$	Weight function for objective 1 to 5
XRD	X-ray diffraction
ZrO_2	Zirconium dioxide
α	Momentum constant
β	Beta
δ	Delta
δ	Delta
η	Learning rate
ω	Omega

Chapter 1
Introduction to Renewable Energy Resources and Sustainable Feedstocks for Biodiesel

1.1 Introduction

Renewable energy sources (RES) and their role in sustainable biodiesel production are of primary concern in addressing the critical issues to meet the global energy demand as an alternative to fossil fuels. The challenges emphasize the need to transition to RES to ensure energy security avoid shortages and reduce environmental impact. The classification of RES shows their potential to provide cleaner and more sustainable energy resources. Detailed insights into different forms of renewable energy resources and their contributions to reducing reliance on fossil fuels are discussed. The importance of biofuel as a sustainable alternative to conventional diesel is highlighted, focusing on its potential to meet stringent energy demands and mitigate environmental pollution. The advantages and limitations of biofuels are discussed including the use of four generations of feedstocks for biodiesel production. The key challenges in selecting sustainable feedstocks, such as balancing food security, environmental impact, and economic viability, are highlighted. General guidelines for choosing biodiesel feedstocks that emphasize sustainability and large-scale availability are proposed. Finally, practical insights into biodiesel production and the factors influencing its sustainability are discussed. The above information is elaborated in subsequent sections.

1.2 Energy Demand and Consumption: Issues and Challenges

Energy is the primary source of every country's or the world's economic growth [1]. Energy demand and consumption are increasing due to the ever-increasing population and modernization [2]. In India, continued urbanization (raising incomes and

M. Patel G. C. et al., *Biofuel Production, Performance, and Emission Optimization*, Springer Optimization and Its Applications 215,
https://doi.org/10.1007/978-3-031-84806-3_1

improved living standards led millions of households to purchase new appliances, air conditioners, and automobiles) and industrialization (to develop structures for buildings and transport vehicles, processing, and packing) led to consuming more energy demand resulting in more energy crises crisis [3, 4]. Figure 1.1 displays the details of trends estimated between 1972 and 2015 towards global energy consumption, viz., various energy resources [5]. Figure 1.1 depicts the intense growth observed with zero greenhouse gas emission energy sources (namely wind and solar) since 2004, but the trajectory of primary energy demand remains incompatible with fossil fuel. The zero-emission approach by 2050 can be accomplished with alternate fuels (hydro, nuclear, solar, wind, biomass, and their diverse combination), but the requirement of huge investments and infrastructure limits energy resource usage [5, 6]. Due to their abundant availability and economy, fossil fuels—coal, oil, and natural gas—account for more than 80% of the energy used worldwide to date [7]. The issue of energy demand can be partly solved with fossil fuel consumption but poses significant environmental and health problems.

The total amount of fossil fuel reserves and gross domestic product reserves by different countries are presented in Fig. 1.2a, b. Figure 1.2 shows that the different country's reserves of fossil fuels in different countries cumulatively constitute 80%. Figure 1.2a shows that coal is the primary constituent fuel consumed by most countries. Coal combustion for energy production poses challenges in two ways: it produces 58% more energy compared to oil when a unit quantity of fuel (i.e. 1 kg of coal and oil) is combusted, and coal fuel emits twice as much carbon dioxide (CO_2) than as natural gas estimated by targeting 1 MJ of energy production. This comparison explains that burning coal generates significant CO_2 emissions [5].

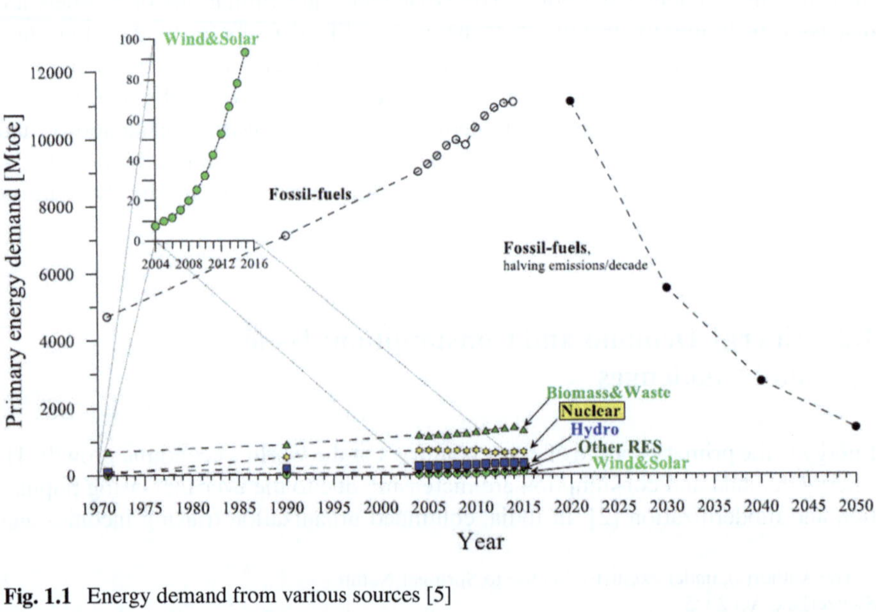

Fig. 1.1 Energy demand from various sources [5]

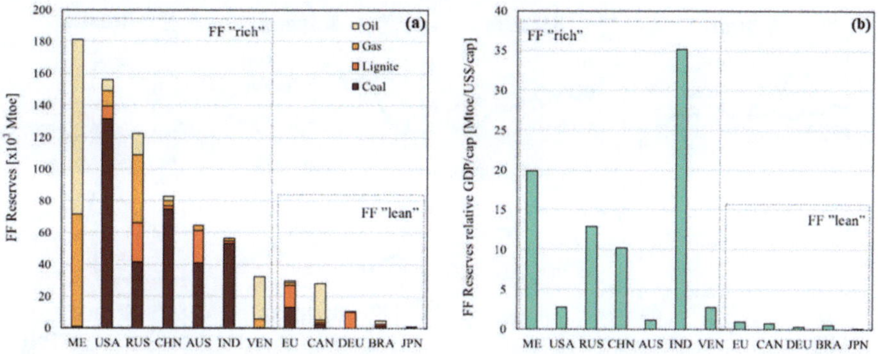

Fig. 1.2 Fossil fuel reserves by different countries: (**a**) in quantity and (**b**) gross domestic product/per capita [5]

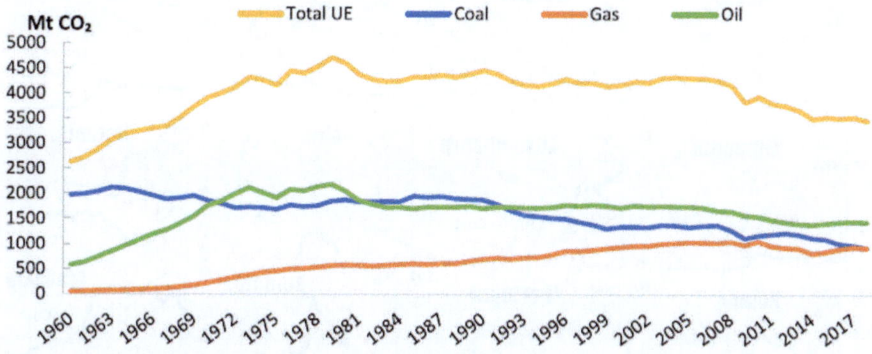

Fig. 1.3 CO_2 emissions from combustible fuels in the European Union: 28 countries [11]

Natural gas has also been shown to have the potential as an alternative fuel, with gas production industries projected to meet global energy demand [8]. Natural gas supply experiences more fluctuations due to seasonal and temperature factors. Challenges in supplying natural gas from production locations to consumption areas include the following [8, 9]: pipelines cannot be placed below sea level (i.e. at depths greater than 100 m), underwater supply is costly and risky, permanent installation fixation of pipelines, pressure variations in gas supply, and seasonal fluctuations in gas pipeline contracts.

Energy production using combustible fossil fuels results in significant emissions, including carbon dioxide, nitrous oxide, and particulate matter, leading to global warming and causing the premature death of millions of people [10]. CO_2 is the primary emission produced by all three combustible fossil fuels. A survey conducted in 28 countries revealed that CO_2 emissions are predominantly from coal 75.59%, followed by oil 22.5%, and gas of 1.91% (refer to Fig. 1.3) [11]. Figure 1.4 illustrates the type of combustible fuel used by 28 European countries based on the availability and favourable conditions for domestic energy resources [5].

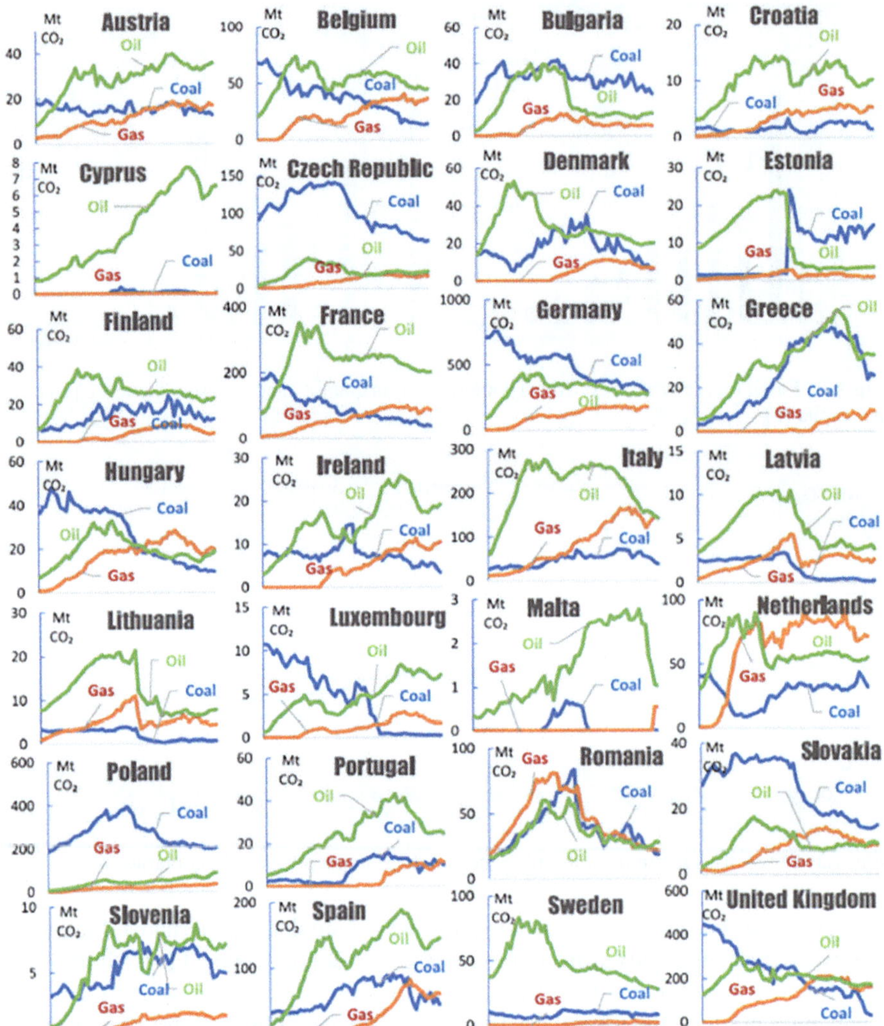

Fig. 1.4 CO_2 emissions by type of combustible fuels of European Union-28 countries [11]

The combustion of crude oil-based fossil fuels has led to meeting energy demand, but it has also posed potential challenges: high market price, negative environmental impact, lack of trust, transparency and funding, political pressure, and abundant availability for oil exploitation. The consumption of petroleum-based raw materials has increased 200 times more in the last century. The fast depletion rate of petroleum products and excessive use in the transportation sector (>57% of total oil consumption) pose problems for petrochemical industries. To mitigate the drawbacks of fossil fuels, it is essential to search for alternative fuels to meet the stringent energy demand.

1.3 Introduction to Renewable Energy Resources (RES) and Their Importance

Fossil fuel-derived non-renewable energy sources currently satisfy 84% of the energy demand [12]. Researchers and scientists have been searching for commercially viable, technically practicable, and environmentally acceptable alternative fuels for practical uses due to pollution, health concerns, and the rapid depletion of fossil fuels [13].

Hydro, wind, solar, geothermal, and biomass are ecologically and economically sustainable RES alternatives to fossil fuels [14]. RES are region-specific, which can be inconsistent and unreliable, requiring demanding new technologies, high capital investments, and infrastructure. For example, to generate energy from the wind, the minimum average wind speed for a turbine to run must be ≥4 m/s [15]. Continuous sunlight and intense daytime striking ensure the solar collector produces maximum energy [16]. Constant water supply throughout the year provides effective hydropower-based energy production [17]. A climatic changes such as floods, drought, and wind storms make solar, wind, and hydropower difficult for energy production [18]. Although the above RES can meet the energy demand, they pose potential challenges or barriers such as cost, policy, and market share. Significant gaps exist in technologies for commercialization and industrialization [19]. Therefore, studying potential feedstocks for renewable energy sources is of industrial relevance.

1.4 Classification of RES

RES can be classified into several types based on their sources and methods of harnessing [20]. The most common types of renewable energy include:

1. **Solar Energy**: Solar energy comes from sunshine, is absorbed by photovoltaic cells, and is used to heat water or produce electricity for various uses.
2. **Wind Energy**: Wind turbines generate electricity from wind power, a major RES.
3. **Hydroelectric Energy**: Hydropower is a form of energy generation that uses flowing water or dams to turn turbines and generate electricity.
4. **Tidal Energy**: Tidal power generates reliable electricity by driving turbine generators using twice-daily tidal currents.
5. **Geothermal Energy**: Geothermal energy utilizes heat from the Earth's core to produce electricity or heat water for various purposes.
6. **Biomass Energy**: By converting organic resources like garbage or plant remnants into power, biomass offers a more sustainable and clean energy source.
7. **Hydrogen**: Using surplus renewable electricity, hydrogen, a renewable secondary by-product, can be created through electrolysis, providing a clean alternative fuel source.

These renewable energy sources are crucial in reducing greenhouse gas emissions, combating climate change, and ensuring a sustainable energy future [21]. Solar energy comes from sunshine, is absorbed by photovoltaic cells, and is used to heat water or produce electricity for various uses.

1.5 Importance of Biofuels and Their Applications

Biofuels play a significant role in addressing global energy needs and have various applications that contribute to sustainability and environmental protection. Here are the key points regarding the importance of biofuels and their applications:

1. **Transportation**: Biofuels are crucial for the transportation sector, which consumes substantial energy globally. They offer a renewable alternative to fossil fuels, reducing emissions and dependence on oil [22].
2. **Energy Generation**: Biofuels can be used in power generation, particularly in backup systems where emissions are a concern. They are beneficial for facilities like schools and hospitals located in residential areas [23].
3. **Heating**: For heating needs, biofuels like bio heat have become more and more popular. They can lower emissions of sulphur dioxide and nitrogen dioxide when used as home heating oil in both residential and commercial boilers [24].
4. **Electronics Charging**: Research has shown that biofuel can generate electricity for charging electronics, potentially replacing batteries in devices like computers and cell phones [25].
5. **Environmental Clean-up**: Biofuels are environmentally friendly and can assist in cleaning up oil spills and grease. They have been tested successfully as cleaning agents for areas contaminated by crude oil [26].
6. **Industrial Applications**: Biofuels can serve as industrial solvents for cleaning metal due to their non-toxic nature. They also have potential applications in paint removal and adhesive products, offering a safer alternative for certain applications [27].
7. **Energy Security**: By reducing the need to import oil, biofuels can help stabilize economies in the long run, minimizing heavy oil import costs and positively impacting overall economic stability [28].
8. **Sustainability**: When compared to traditional fuels, the development of biofuels from a variety of feedstock's provides a sustainable energy alternative that can help mitigate climate change by reducing greenhouse gas emissions [29].

In conclusion, biofuels are critical for lowering greenhouse gas emissions, improving energy security, and offering long-term substitutes for fossil fuels in a variety of industries, including heating, energy production, transportation, and industrial uses. Their importance lies in their ability to address environmental concerns while meeting global energy demands sustainably.

1.6 Need for Biofuel Production to Meet Global Energy Demand

The need for biofuel production is driven by the global energy demand, particularly in the transport sector. Biofuels help reduce dependence on unpredictable foreign supplies, exhaustible resource depletion, and conventional and greenhouse gas emissions [30]. The demand for biofuels is expected to increase, with global biofuel demand expanding significantly over the next 5 years to meet climate and energy security goals [31]. Liquid biofuels are projected to account for 14% of the world's transportation energy consumption by 2050, up from 4% in 2020, highlighting the growing significance of biofuels in the world's energy mix [32].

The two main biofuels produced worldwide are bioethanol and biodiesel, with ethanol accounting for over 70% of total biofuel production [32]. Fossil fuels can be replaced with RES such as animal fats, sugar crops, starch crops, oilseed crops, and animal fats. Furthermore, when compared to conventional fuels, second and third generation biofuels have the potential to reduce greenhouse gas emissions significantly [30].

Additionally, biofuel production can lead to economic benefits such as increased farm income, reduced fossil fuel imports, and lower vulnerability to supply disruptions. However, biofuel development may also have negative effects, such as different land use patterns, depletion of water supplies, pollution of the air and water, and higher food prices [30]. Balancing the benefits and challenges of biofuel production is essential to ensure sustainable energy practices and mitigate environmental impacts while meeting the growing global energy demand.

1.6.1 Advantages and Limitations of Biodiesels

RES meet 14% of global energy demand, with biomass contributing approximately 70% [33, 34]. Many academic publications have stated that the terms "biofuel" and "biomass" have the same meaning: biofuel is the by-product of processing biomass, whereas biomass is the raw material [19, 35]. RES-based biodiesel or biofuel is a non-fossil fuel-derived alternative to Petro-diesel derived from plant or animal biomass [36]. Numerous advantages enable biofuels to be competitive with to fossil fuels (in particular, diesel), as follows [19, 37]:

1. Biofuel production and their mix with traditional fossil fuels meet standards can be used in internal combustion engines without significant modifications to diesel engines.
2. Biofuels significantly benefit industries and societies, directly impacting the environment and economy.
3. Significant by-products are derived that are suitable for distinguished applications and reduce fuel costs during biofuel production.

4. Biofuels reduce greenhouse gas emissions (in particular CO_2). However, emissions depend on the starting biomass or feedstock and the production route employed.
5. Biofuel reduces pollutants released into the atmosphere, namely SOx, NOx, and CO.
6. Biofuels burn cleaner and more efficiently than petroleum-based alternatives.
7. Biofuels serve as the basis for energy security and minimize the potential scarcity risk.
8. Biofuel production offers jobs that could improve societal status and the economy.
9. Biofuel decreases dependence on crude oil in neighbouring countries.
10. Biofuels are often biodegradable and emit less CO_2, SO_2, and hydrocarbons.
11. The physical properties of biofuel, namely high flash point temperature, lubricating properties, and octane number, make it being a potential competitor to fossil fuel.

The challenges involved in taking significant benefits of biodiesel are as follows [19, 37]:

1. Global competition.
2. The selection of cost-effective potential feedstocks requires intense studies to search for new biodiesel production chains to increase the availability and use of biofuels.
3. Issues concerning the food versus fuel war, i.e. most biodiesel productions are accomplished with edible oils (vegetable or plant-derived oil), resulting in the risk of diverting agricultural land at the cost of food supply on a global scale.
4. Plant-based feedstock's face issues with land and deforestation.
5. Biodiesel produces 8% less energy per gallon than conventional diesel.
6. Biodiesel is comparatively higher in density than typical diesel.
7. Biodiesel possesses less favourable cold flow properties (can freeze or gel and clog filters resulting in higher viscosity to be pumped from the fuel storage tank to the engine).
8. Biodiesel contains polyunsaturated fatty acids that affect oxidation stability.
9. Biodiesel emits higher NOx than diesel.
10. Copper and brass undergo corrosion against biodiesel.
11. Biodiesel causes higher engine wear.

1.6.2 Feedstock's for Biodiesel Production

The total price of biodiesel depends primarily on feedstock availability and cost [38]. The survey concluded that approximately 70–95% of the biodiesel price is accountable to feedstock alone [39]. Therefore, the selection of feedstock's also relies on region-specific availability and economic-specific relevance to industries

[40]. Biofuels are derived from various sources such as living organisms, biomass (firewood, methane, and petroleum), animals, plants, biological materials, fats and oils, and waste feedstock, based on different evolutions/generations, highlighting their advantages and disadvantages [41]. In addition, the probable reasons for the transition shift from Generation 1 → Generation 2 → Generation 3 → Generation 4 are explained below:

1. The increased cost of food and food-based products, lack of availability of land for cultivation (food vs. fuel war), and high cost are a few reasons for Generation 1.
2. The requirement of high alcohol content and non-edible crops on farming land directly influence society and food production, resulting in few observations with Generation 2.
3. A lack of efficient technologies for oil extraction from feedstock's, low lipid content, harvesting, and processing difficulties are observed in Generation 3.

1.6.3 Generation of Feedstock's for Biodiesel Production Highlighting Their Advantages and Limitations

Generating feedstock's for biodiesel production involves utilizing various sources such as oils (vegetable, recycled cooking) and animal fats. These feedstock's offer distinct qualities that impact the properties of biodiesel. In the USA, soybean oil and canola oil are commonly used feedstock's, with soybean oil significantly contributing to biodiesel production. Additionally, recycled cooking oils like yellow grease and animal fats such as tallow and white grease are utilized for biofuel production. The feedstock selection influences the properties of biodiesel, such as cloud point, cetane number (CN), and oxidative stability, which are vital for engine performance and storage concerns [42].

Various catalysts are utilized during transesterification to transform these feedstock's into biodiesel effectively. When feedstock's have a low free fatty acid (FFA) concentration, alkali catalysts are used; acid catalysts are favoured for feedstocks with a greater FFA content. Transesterification involves reacting fats or oils with alcohol to form biodiesel esters and glycerol. This process is essential for overcoming challenges associated with using straight vegetable oils directly as fuel due to issues like poor atomization, incomplete combustion, and high viscosity. For biodiesel production to be sustainable, the choice of feedstock is essential, considering variables like calorific value, moisture content, oil content, and contaminants [43].

The generation of feedstocks for biodiesel production involves carefully selecting oils (vegetable and waste cooking) and animal fats based on their properties and suitability for transesterification processes to produce biodiesel efficiently while meeting industry standards and environmental considerations. Table 1.1 shows the generation of feedstocks with their advantages and disadvantages.

Table 1.1 Generation of feedstocks with their advantages and disadvantages

Generation	Feedstocks	Advantages [41, 44]	Disadvantages [41, 44]
First generation [Edible]	Pistachio [45], Mustard [46], Rapeseed [47], Soybean [48], Sunflower [49], Maize [50], Potato [51], Rice bran oil [52], Coconut oil [53], Olive oil [54], Radish [55], Palm [56], Tiger nut [57], Walnut [58]	• Ease of biodiesel conversion • Ease of availability of feedstocks • Biodegradability • Environmental and societal benefits • Energy security	• Competition in land use and low crop yield • Blending required with diesel • High carbon footprint • Highest food price • Negative impact on biodiversity
Second [Non-edible or non-food crop]	Candlenut [59], Cotton seed [60], *Calophyllum inophyllum* [61], Castor [62], Pongamia [63], Polanga [64], *Jatropha curcas* [65], Jojoba [66], Karanja [67], Mahua (Madhuca indica) [68], Milk bush [69], Nag champa [70], Neem [71], Kusum [72], Rubber seed [73], Soapnut [74]	• No impact on food supply • Feedstocks grow in non-fertile land • Low production and feedstock cost • Use complete plant rather than only seeds and residues • Higher yield with limited land • Feedstocks are processed similarly to existing infrastructure • Energy security • Environmental and societal benefits	• Low yield for some feedstocks • Agriculture and forest residue use decreases soil quality, causing erosion • The production process is comparatively complex • Lack of effective technologies for commercialization
Third [Waste oils]	Animal tallow [75], Biomass pyrolysis [76], *Botryococcus braunii* [77], Chicken fat [78], Chlorella vulgaris algae [79], Dunaliella salina algae [80], Poultry fat [81], Fish [82], Waste cooking oil [83]	• No food and land competition. • Waste oil for biofuel production • Wastewater or sea water for higher algae growth rate • No toxic compounds with high yield • Low or no-cost feedstock • Enhance performances with a mix of first and second generation feedstocks	• High energy consumption for algae cultivation • High cost of algae-oil extraction • Lack of efficient technologies for biofuel production • Harvesting and processing difficulty • Higher production cost • Low lipid content
Fourth [Solar]	Photobiological solar [44]	• Higher lipid content and energy • High growth rate • High CO_2 absorbing capacity	• Research in development stage rather than commercialization • High cost and lack of studies

1.6.4 Challenges in Selection of Sustainable Feedstocks for Biodiesel Production

Feedstocks are selected for biofuel production based on their practical usefulness to engine applications and economic relevance as follows:

1. Feedstocks must ensure sustainability (social, economic, and environmental) and not generate negative ecological effects [84].
2. Potential and sustainable feedstock for energy production needs to be identified in aquatic or soil environments. They should not have a negative environment when meeting energy demand [85].
3. Feedstocks must be abundantly available, low cost, easy to convert from oil to fuel, and have biofuel properties similar to neat diesel [82].
4. The selected feedstock should have a high oil fraction in the seed, leading to reduced biofuel prices as approximately 70–95% of total fuel costs are determined by the feedstock alone.
5. The selected feedstocks should allow for various additional uses such as pharmaceutical, curries, and painting [86].
6. Biological and physicochemical characteristics evaluation of feedstocks useful for biofuel production [87, 88].

India is a densely populated country with the second-largest population globally [89]. Feedstocks rely on abundant availability at the nearest geographical locations, existing infrastructure, crops, and crude oil prices [90]. In addition, feedstocks grown on agricultural land are not economically feasible due to the huge food grain requirements for people's livelihoods [91]. This situation hinders biofuel conversion from seed oil for commercialization. Distinguished researchers must pay significant attention to identifying locally available, novel and, cost-effective feedstocks [92]. Indeed, an energy crisis is a global issue, and respective governments do not restrict harvesting fruits or seeds grown on forest lands useful for biofuel production. Instead, the government must subsidize farmers who grow these crops on non-fertile land [93].

1.6.5 Challenges in Commercialization and Industrialization of Renewable Energy Sources

Some important existing laws and regulations have an impact on renewable energy trading and operations:

- Identifying useful and efficient biodiesel feedstocks and their technological transformations without affecting the environment and economy is still a challenge for researchers [94].

- Each biofuel generation feedstock possesses significant challenges, such as:
  - First-generation feedstock must compete with food demand, leading to increased food and animal feed prices [95].
  - Second-generation feedstock uses feedstock derived from waste resources, leading to higher energy efficiency. However, this must compensate for the slow and low degradability of lignocellulose and the large areas needed [96].
  - Microalgae cultivation at a large scale requires a significant amount of water, nitrogen, and phosphorus for third-generation biofuel production to cater to industry demand [97]. In addition, algae growth relies on fertilizers, which consume more energy and generate more greenhouse gas emissions than biofuel [97, 98].
  - Fourth-generation feedstocks and their biodiesel production techniques are developing and need to be industry-ready for scaling up energy production [90, 99].
  - Low conversion efficiencies with second—and third-generation biofuels require significant investments in optimizing technologies to make them cost-effective [94].
- The main challenges identified for efficient production of biodiesel to meet global energy needs are [100]:
  - Need for extra land.
  - Pressure on water resources for feedstock.
  - Advanced conversion technologies for cleaner and smoother production.
  - Most conversion technologies currently employed use organic chemicals that pollute the environment.
  - Ensuring the stability and durability of engines fuelled with biodiesel without modification is a significant challenge.
- The efficient and timely availability of feedstock for biodiesel production in the industry poses significant challenges to commercialization [101].
- The challenges in commercialization faced by industries are discussed below:
  - The availability of feedstocks is truly reliant on geographical locations and climatic conditions [90]. Furthermore, variations in feedstock quality and seasonal availability pose challenges to scaling up production at the industry scale [102].
  - The need for efficient and cost-effective conversion technologies for biofuel production from second-, third-, and fourth-generation feedstocks remains a significant challenge [103, 104].
  - The scalability of biofuel production towards commercialization from a laboratory scale requires significant attention to ensure process stability and efficiency [105, 106].
  - Effective management of by-products generated from biofuel production without affecting environmental impacts and process economics is challenging [107].

- Industry-scale biofuel production with feedstocks grown on agricultural land raises food security concerns and deforestation [91].
- The present biofuel conversion technologies are water- and energy-intensive, significantly raising concern for the industry in water-scarce geographical locations [108].
- Governments must establish policies and incentives that include subsidies and tax incentives for resources (growing feedstock, fertilizers, water, land, transport), advanced biofuel conversion technologies that are beneficial to promote biofuel production [109].
- Meeting stringent governmental and environmental regulations for biofuel production, usage, and distribution remains complex and time-consuming [110, 111].
- Building public awareness about feedstocks, revenue, biofuel usage, production, and market penetration remains a challenging task for the government.

1.6.6 Salient Points to Consider for the Selection of Feedstocks for Biodiesel Production

Feedstocks selected for biodiesel production should ensure sustainability (socioeconomic and environmental benefits) for and are of industrial relevance [84]. Methods for searching for feedstocks in the environment (aquatic and soil) to produce biofuels that are not harmful to the environment or meet energy needs have been developed [85]. The choice of feedstock raw materials is region-specific and depends on the available resources at geographic locations [112]. In densely populated countries like India, where food security and affordability are crucial, first-generation feedstocks are not recommended [91, 113]. Fourth-generation feedstocks and their biodiesel production techniques are currently in the development stage and need to be industry-ready for scaling up energy production [90, 99]. Non-food crop-based, non-edible seeds (containing approximately 37–50% oil) predominantly grown in India and Ethiopia [114]. However, non-edible seeds are less common in other regions such as Asia, America, and Africa. Therefore, selecting second-generation feedstocks for biodiesel production focuses on region-specific low-cost feedstock and biofuel production costs. Table 1.2 shows the secondary feedstock for biodiesel conversion. Researchers across the globe use diverse feedstocks derived from plant and seed oils for biofuel production. Each feedstock differs, with oil content ranging from 17% for cotton seed to 75% for polenta. Therefore, the choice of feedstocks for biodiesel conversion is based on their abundant availability in specific geographical locations and oil content, which ensures lower production costs. The methods used for oil conversion from diverse feedstocks are mechanical pressing, solvent extraction, supercritical CO_2 extraction, and enzymatic extraction. Transesterification is the most commonly used technique for biodiesel conversion from crude oil. The variation in biodiesel conversion from oil is due to the influenced transesterification parameters and catalyst used. Catalyst selection is crucial for achieving maximum yield. NaOH and KOH catalysts are widely accepted catalysts for biodiesel conversion. Raw materials with high seed oil content are preferred for biodiesel production. Recall that feed costs usually constitute 70–95% of the final cost of biodiesel. The performance of biodiesel yield is influenced by the optimization of parameters and biodiesel extraction methods.

Table 1.2 Summary of results of second-generation feedstocks for biodiesel conversion

Feedstock	Oil fraction	Grown in region	Method of oil extraction	Method of oil conversion	Yield (%)	Refs.
Candlenut	30–60%	Indo-Malaysia	Mechanical press method, Soxhlet extraction technology, and supercritical CO_2 extraction	Transesterification	NR	[59]
Cotton seed	17–25%	China, India, USA, Brazil, Pakistan	NR	Microwave-assisted transesterification	95% for KOH and <91% for CaO catalyst	[60]
Polanga (*Calophyllum inophyllum*)	75%	Malaysia, Indonesia, India Australia, and East Africa	NR	Transesterification (KOH catalyst)	95.8%	[61, 64]
Castor	45–50%	NR	NR	Transesterification	NR	[62]
Pongamia	30–40%	NR	NR	Transesterification	NR	[63]
Jatropha curcas	20–60%	Nigeria, India, Indonesia, Malaysia, Thailand, etc.	Aqueous enzymatic extractions: 33.44%, ultrasound-assisted solvent extraction: 10–59%, supercritical oil extraction: 32–60%, solvent extraction methods: 31–57% oil, mechanical extraction: 25–30% oil	Transesterification	95–100% for NaOH, 96.1 for CaO	[65]

(continued)

Table 1.2 (continued)

Feedstock	Oil fraction	Grown in region	Method of oil extraction	Method of oil conversion	Yield (%)	Refs.
Jojoba	40–50%	Deserts: the USA, Israel, Mexico, India	NR	Transesterification	93% for KOH	[66]
Karanja	25–40%	NR	Mechanical extraction	Transesterification (KOH)	91%	[115]
Jatropha	20–60%	NR	Mechanical extraction	Transesterification (KOH)	93%	[115]
Mahua (Madhuca indica)	35–40%	Southern Asia	NR	Transesterification (KOH)	88.7%	[116]
Milk bush	41.2%	Nigeria	Mechanical extraction	Transesterification (KOH and CaO)	94.33%	[117]
Nag champa	65–70%	India	Mechanical extraction	Ultrasound transesterification (H_2SO_4)	79%	[118]
Neem	20–30%	Nigeria, India, Argentina, China, Brazil, Egypt	NR	Transesterification (H_2SO_4)	85.13%	[119]
Kusum	30–40%	India, China, Timor, Sri Lanka, and Myanmar	NR	Transesterification (KOH)	98.48%	[72]
Rubber seed	NR	India, Thailand, Vietnam, Indonesia, China	NR	Transesterification (CaO catalyst)	97.84%	[120]
Soapnut	NR	India	Solvent extraction	Transesterification coal burnt residue	89%	[121]

NR = not recorded, KOH = potassium hydroxide, CaO = calcium oxide, NaOH = sodium hydroxide, H_2SO_4 = sulphuric acid

1.7 Concluding Remarks

This chapter discusses the global energy demand and highlights the need for renewable energy resources and sustainable feedstocks for biodiesel. The following are the salient points drawn:

1. Increased global energy demand is a result of the rapid rise in population and energy consumption to cater for the needs of industrial and transportation sectors. The reliance on fossil fuels, which comprise over 80% of global energy

consumption, presents challenges, including environmental degradation, health issues, and unsustainable depletion rates.
2. RES such as hydro, wind, solar, and biomass have shown the importance of mitigating the adverse effects of fossil fuels. Biodiesel is a proven sustainable alternative that reduces greenhouse gas emissions, improves energy security, and contributes to the country's economic growth through local production, reducing reliance on oil imports.
3. The feedstock alone accounts for 70–95% of the total biodiesel production cost. Therefore, the selection of sustainable feedstocks (after studying different generations) for biodiesel production was discussed, along with their advantages and limitations. Second-generation feedstocks were found to have sustainable, technical, and economic benefits.
4. The selection of feedstocks should be based on the following criteria: grown in non-arable land, does not cause negative environmental effects, low cost, ease of oil to biodiesel conversion, maximum oil fractions, and generate revenue for farmers.
5. The feedstock for biodiesel conversion must be selected based on high oil content and abundant availability at specific geographic locations. The mechanical pressing followed by solvent extraction methods is the preferred choice for oil extraction. The appropriate selection of transesterification parameters and catalysts is of industrial relevance to maximize biodiesel conversion and lower the production cost.

References

1. Foxon, T.J.: Energy and Economic Growth: Why we need a new pathway to prosperity. In: Routledge Studies in Energy Transitions. Routledge (2017)
2. Asif, M., Muneer, T.: Energy supply, its demand and security issues for developed and emerging economies. Renew. Sust. Energ. Rev. **11**(7), 1388–1413 (2007)
3. Franco, S., Mandla, V.R., Rao, K.R.M.: Urbanization, energy consumption and emissions in the Indian context: a review. Renew. Sustain. Energy Rev. **71**, 898–907 (2017)
4. Shahbaz, M., Lean, H.H.: Does financial development increase energy consumption? The role of industrialization and urbanization in Tunisia. Energy Policy. **40**, 473–479 (2012)
5. Johnsson, F., Kjärstad, J., Rootzén, J.: The threat to climate change mitigation posed by the abundance of fossil fuels. Clim. Pol. **19**(2), 258–274 (2019)
6. Kleijn, R., et al.: Metal requirements of low-carbon power generation. Energy. **36**(9), 5640–5648 (2011)
7. Holechek, J.L., et al.: A global assessment: can renewable energy replace fossil fuels by 2050? Sustain. For. **14**(8), 4792 (2022)
8. Kumar, S., et al.: Current status and future projections of LNG demand and supplies: a global prospective. Energy Policy. **39**(7), 4097–4104 (2011)
9. Keyaerts, N., et al.: Gas market distorting effects of imbalanced gas balancing rules: Inefficient regulation of pipeline flexibility. Energy Policy. **39**(2), 865–876 (2011)
10. Ogunkunle, O., Ahmed, N.A.: Overview of biodiesel combustion in mitigating the adverse impacts of engine emissions on the sustainable human–environment scenario. Sustain. For. **13**(10), 5465 (2021)

11. Paraschiv, S., Paraschiv, L.S.: Trends of carbon dioxide (CO2) emissions from fossil fuels combustion (coal, gas and oil) in the EU member states from 1960 to 2018. Energy Rep. **6**, 237–242 (2020)
12. Opeyemi, B.M.: Path to sustainable energy consumption: The possibility of substituting renewable energy for non-renewable energy. Energy. **228**, 120519 (2021)
13. Perumal, V., Ilangkumaran, M.: Experimental analysis of operating characteristics of a direct injection diesel engine fuelled with Cleome viscosa biodiesel. Fuel. **224**, 379–387 (2018)
14. Salvarli, M.S., Salvarli, H.: For sustainable development: future trends in renewable energy and enabling technologies. In: Renewable Energy-Resources, Challenges and Applications. IntechOpen (2020)
15. Islam, M., Saidur, R., Rahim, N.: Assessment of wind energy potentiality at Kudat and Labuan, *Malaysia using Weibull distribution function*. Energy. **36**(2), 985–992 (2011)
16. Sabiha, M., et al.: Progress and latest developments of evacuated tube solar collectors. Renew. Sustain. Energy Rev. **51**, 1038–1054 (2015)
17. Oberle, P., et al.: Adapted hydropower-driven water supply system: assessment of an underground application in an Indonesian karst area. Appl Water Sci. **6**, 259–278 (2016)
18. Khan, I., Alam, F., Alam, Q.: The global climate change and its effect on power generation in Bangladesh. Energy Policy. **61**, 1460–1470 (2013)
19. Al Qubeissi, M., El-Kharouf, A.: Renewable energy: resources, challenges and applications. (2020)
20. Ellabban, O., Abu-Rub, H., Blaabjerg, F.: Renewable energy resources: Current status, future prospects and their enabling technology. Renew. Sustain. Energy Rev. **39**, 748–764 (2014)
21. Androniceanu, A., Sabie, O.M.: Overview of green energy as a real strategic option for sustainable development. Energies. **15**(22), 8573 (2022)
22. Nogueira, L.A.H., et al.: Biofuels for transport. In: Future Energy, pp. 173–197. Elsevier (2020)
23. Nikkhah, A., et al.: Comparative review of three approaches to biofuel production from energy crops as feedstock in a developing country. Bioresour. Technol. Rep. **10**, 100412 (2020)
24. Bazooyar, B., Ghorbani, A., Shariati, A.: Combustion performance and emissions of petrodiesel and biodiesels based on various vegetable oils in a semi industrial boiler. Fuel. **90**(10), 3078–3092 (2011)
25. Choi, S.: Biofuel cells and biobatteries: misconceptions, opportunities, and challenges. Batteries. **9**(2), 119 (2023)
26. Negi, H., Singh, R.K.: A review on lignin utilization in petroleum exploration, petroleum products formulation, bio-fuel production, and oil spill clean-up. In: Biomass Conversion Biorefinery, pp. 1–12 (2020)
27. Hu, J., et al.: Study on the solvent power of a new green solvent: biodiesel. Ind. Eng. Chem. Res. **43**(24), 7928–7931 (2004)
28. Kumar, S., Shrestha, P., Salam, P.A.: A review of biofuel policies in the major biofuel producing countries of ASEAN: production, targets, policy drivers and impacts. Renew. Sustain. Energy Rev. **26**, 822–836 (2013)
29. Moustakas, K., et al.: A review of recent developments in renewable and sustainable energy systems: key challenges and future perspective. Renew. Sustain. Energy Rev. **119**, 109418 (2020)
30. EPA. https://www.epa.gov/environmental-economics/economics-biofuels
31. IEA. Renewables 2022. https://www.iea.org/reports/renewables-2022/transport-biofuels (2022)
32. Masi, M., et al.: Roadmap to 2050: the Land-water-energy nexus of biofuels, in roadmap to 2050 the land-water-energy nexus of biofuels. In: Sustainable Development Solutions Network and Fondazione Eni Enrico Mattei, pp. 1–164 (2021)
33. Popp, J., et al.: Bioeconomy: Biomass and biomass-based energy supply and demand. New Biotechnol. **60**, 76–84 (2021)
34. Tian, J., et al.: Global low-carbon energy transition in the post-COVID-19 era. Appl. Energy. **307**, 118205 (2022)

35. Clauser, N.M., et al.: Biomass waste as sustainable raw material for energy and fuels. Sustain. For. **13**(2), 794 (2021)
36. Demirbas, A.: Biofuels from agricultural biomass. Energy Sources Part A. **31**(17), 1573–1582 (2009)
37. Srinivas, D., Satyarthi, J.K.: Challenges and opportunities in biofuels production. (2012)
38. Maheshwari, P., et al.: A review on latest trends in cleaner biodiesel production: role of feedstock, production methods, and catalysts. J. Clean. Prod. **355**, 131588 (2022)
39. Bhuiya, M., et al.: Prospects of 2nd generation biodiesel as a sustainable fuel—Part: 1 selection of feedstocks, oil extraction techniques and conversion technologies. Renew. Sust. Energ. Rev. **55**, 1109–1128 (2016)
40. de Jong, S., et al.: Cost optimization of biofuel production–The impact of scale, integration, transport and supply chain configurations. Appl. Energy. **195**, 1055–1070 (2017)
41. Pandey, A.: Biofuels: alternative feedstocks and conversion processes. Academic Press (2011)
42. Zulqarnain, et al.: Overview of feedstocks for sustainable biodiesel production and implementation of the biodiesel program in Pakistan. ACS Omega. **6**(29), 19099–19114 (2021)
43. Elgharbawy, A.S., et al.: A review on biodiesel feedstocks and production technologies. J. Chil. Chem. Soc. **66**(1), 5098–5109 (2021)
44. Singh, D., et al.: A review on feedstocks, production processes, and yield for different generations of biodiesel. Fuel. **262**, 116553 (2020)
45. Göncü, B., Gülşen, H., Hoşgün, E.Z.: Bioethanol production from pistachio (pistacia vera L.) shells applying ozone pretreatment and subsequent enzymatic hydrolysis. Environ. Technol. **42**(15), 2438–2446 (2021)
46. Rezania, S., et al.: Biodiesel production from wild mustard (Sinapis Arvensis) seed oil using a novel heterogeneous catalyst of LaTiO3 nanoparticles. Fuel. **307**, 121759 (2022)
47. Gupta, R., et al.: Life cycle assessment of biodiesel production from rapeseed oil: influence of process parameters and scale. Bioresour. Technol. **360**, 127532 (2022)
48. da Silva César, A., et al.: Competitiveness analysis of "social soybeans" in biodiesel production in Brazil. Renew. Energy. **133**, 1147–1157 (2019)
49. Elkelawy, M., et al.: Experimental studies on the biodiesel production parameters optimization of sunflower and soybean oil mixture and DI engine combustion, performance, and emission analysis fueled with diesel/biodiesel blends. Fuel. **255**, 115791 (2019)
50. Choudhary, M., et al.: Enabling technologies for utilization of maize as a bioenergy feedstock. Biofuels Bioprod. Biorefin. **14**(2), 402–416 (2020)
51. Daimary, N., et al.: Potato peels as a sustainable source for biochar, bio-oil and a green heterogeneous catalyst for biodiesel production. J. Environ. Chem. Eng. **10**(1), 107108 (2022)
52. Hoang, A.T., et al.: Rice bran oil-based biodiesel as a promising renewable fuel alternative to petrodiesel: a review. Renew. Sustain. Energy Rev. **135**, 110204 (2021)
53. Dharmegowda, I.Y., et al.: MgO nano-catalyzed biodiesel production from waste coconut oil and fish oil using response surface methodology and grasshopper optimization. Sustain. For. **14**(18), 11132 (2022)
54. Khan, I.W., et al.: Biodiesel production by valorizing waste non-edible wild olive oil using heterogeneous base catalyst: Process optimization and cost estimation. Fuel. **320**, 123828 (2022)
55. Eldiehy, K.S., et al.: Radish (Raphanus sativus L.) leaves: A novel source for a highly efficient heterogeneous base catalyst for biodiesel production using waste soybean cooking oil and Scenedesmus obliquus oil. Renew. Energy. **191**, 888–901 (2022)
56. Wahyono, Y., et al.: Assessing the environmental performance of palm oil biodiesel production in indonesia: A life cycle assessment approach. Energies. **13**(12), 3248 (2020)
57. Tian, Y., et al.: Lewis acid-catalyzed in situ transesterification/esterification of tigernut in sub/supercritical ethanol: An optimization study. Fuel. **245**, 96–104 (2019)
58. Miladinović, M.R., et al.: Valorization of walnut shell ash as a catalyst for biodiesel production. Renew. Energy. **147**, 1033–1043 (2020)
59. Shaah, M.A., et al.: Candlenut oil: review on oil properties and future liquid biofuel prospects. Int. J. Energy Res. **45**(12), 17057–17079 (2021)

60. Sharma, A., Kodgire, P., Kachhwaha, S.S.: Biodiesel production from waste cotton-seed cooking oil using microwave-assisted transesterification: optimization and kinetic modeling. Renew. Sustain. Energy Rev. **116**, 109394 (2019)
61. Ramesh, A., et al.: Influence of hexanol as additive with Calophyllum Inophyllum biodiesel for CI engine applications. Fuel. **249**, 472–485 (2019)
62. Attia, A.M., et al.: The influence of castor biodiesel blending ratio on engine performance including the determined diesel particulate matters composition. Energy. **239**, 121951 (2022)
63. Dhanasekar, K., et al.: A facile preparation, performance and emission analysis of pongamia oil based novel biodiesel in diesel engine with CeO2: Gd nanoparticles. Fuel. **255**, 115756 (2019)
64. Venu, H., Appavu, P.: Al2O3 nano additives blended Polanga biodiesel as a potential alternative fuel for existing unmodified DI diesel engine. Fuel. **279**, 118518 (2020)
65. Ewunie, G.A., et al.: Factors affecting the potential of Jatropha curcas for sustainable biodiesel production: a critical review. Renew. Sustain. Energy Rev. **137**, 110500 (2021)
66. El-Zoheiry, R.M., et al.: Combustion and emission characteristics of Jojoba biodiesel-jet A1 mixtures applying a lean premixed pre-vaporized combustion techniques: An experimental investigation. Renew. Energy. **162**, 2227–2245 (2020)
67. Shrivastava, K., Thipse, S., Patil, I.: Optimization of diesel engine performance and emission parameters of Karanja biodiesel-ethanol-diesel blends at optimized operating conditions. Fuel. **293**, 120451 (2021)
68. Sayyed, S., Das, R.K., Kulkarni, K.: Experimental investigation for evaluating the performance and emission characteristics of DICI engine fueled with dual biodiesel-diesel blends of Jatropha, Karanja, Mahua, and Neem. Energy. **238**, 121787 (2022)
69. Sivamani, S., et al.: Process modelling and simulation of biodiesel synthesis reaction for non-edible yellow oleander (yellow bells) oil and waste chicken fat. Bioenergy Research: Revisiting Latest Development, pp. 129–160 (2021)
70. Nayak, S.K., Mishra, P.C.: Application of Nagchampa biodiesel and rice husk gas as fuel. Energy Sourc. Part A Recov. Utili. Environ. Eff. **38**(14), 2024–2030 (2016)
71. Karmakar, A., Karmakar, S., Mukherjee, S.: Biodiesel production from neem towards feedstock diversification: Indian perspective. Renew. Sustain. Energy Rev. **16**(1), 1050–1060 (2012)
72. Pali, H.S., et al.: Biodiesel yield and properties optimization from Kusum oil by RSM. Fuel. **291**, 120218 (2021)
73. Trirahayu, D.A., et al.: Process simulation and design considerations for biodiesel production from rubber seed oil. Fuels. **3**(4), 563–579 (2022)
74. Chhetri, A., Watts, K.: Surface tensions of petro-diesel, canola, jatropha and soapnut biodiesel fuels at elevated temperatures and pressures. Fuel. **104**, 704–710 (2013)
75. Andreo-Martinez, P., et al.: Waste animal fats as feedstock for biodiesel production using non-catalytic supercritical alcohol transesterification: a perspective by the PRISMA methodology. Energy Sustain. Dev. **69**, 150–163 (2022)
76. Gea, S., et al.: A comprehensive review of experimental parameters in bio-oil upgrading from pyrolysis of biomass to biofuel through catalytic hydrodeoxygenation. Bioenergy Res. **16**(1), 325–347 (2023)
77. Thurakit, T., et al.: High-efficiency production of biomass and biofuel under two-stage cultivation of a stable microalga Botryococcus braunii mutant generated by ethyl methanesulfonate-induced mutation. Renew. Energy. **198**, 176–188 (2022)
78. Chizoo, E., et al.: Adaptive neuro-fuzzy inference system-genetic algorithm versus response surface methodology-desirability function algorithm modelling and optimization of biodiesel synthesis from waste chicken fat. Taiwan Inst. Chem. Eng. **136**, 104389 (2022)
79. Sangeetha, M., et al.: Vibration, acoustic and emission characteristics of the chlorella vulgaris microalgae oil in compression ignition engine to mitigate environmental pollution. Chemosphere. **293**, 133475 (2022)

80. de Souza Celente, G., et al.: Potential use of microalga Dunaliella salina for bioproducts with industrial relevance. Biomass Bioenergy. **167**, 106647 (2022)
81. Santhosh, N., et al.: Poultry fat biodiesel as a fuel substitute in diesel-ethanol blends for DI-CI engine: experimental, modeling and optimization. Energy. **270**, 126826 (2023)
82. Dharmegowda, I.Y., et al.: Optimization for waste coconut and fish oil derived biodiesel with MgO nanoparticle blend: Grey relational analysis, grey wolf optimization, driving training based optimization and election based optimization algorithm. Fuel. **338**, 127249 (2023)
83. Pali, H.S., et al.: Enhancement of combustion characteristics of waste cooking oil biodiesel using TiO2 nanofluid blends through RSM. Fuel. **331**, 125681 (2023)
84. Bhat, M., Sofi, A., Sajith, S.: Does biofuel consumption improve environmental quality? An econometric analysis. Int. J. Environ. Sci. Technol. **20**(2), 2063–2072 (2023)
85. Malode, S.J., et al.: Recent advances and viability in biofuel production. Energy Convers. Manag. X. **10**, 100070 (2021)
86. Ghatak, H.R.: Biorefineries from the perspective of sustainability: feedstocks, products, and processes. Renew. Sustain. Energy Rev. **15**(8), 4042–4052 (2011)
87. Shaah, M.A.H., et al.: A review on non-edible oil as a potential feedstock for biodiesel: physicochemical properties and production technologies. Int. J. Energy Res. **11**(40), 25018–25037 (2021)
88. Singh, D., et al.: A comprehensive review of physicochemical properties, production process, performance and emissions characteristics of 2nd generation biodiesel feedstock: Jatropha curcas. Fuel. **285**, 119110 (2021)
89. Sharma, S.D.: India's fight against the COVID-19 pandemic: lessons and the way forward. India Quart. **78**(1), 9–27 (2022)
90. Sarwer, A., et al.: Suitability of biofuels production on commercial scale from various feedstocks: a critical review. ChemBioEng Rev. **9**(5), 423–441 (2022)
91. Renzaho, A.M., Kamara, J.K., Toole, M.: Biofuel production and its impact on food security in low and middle income countries: Implications for the post-2015 sustainable development goals. Renew. Sustain. Energy Rev. **78**, 503–516 (2017)
92. Lim, S., Teong, L.K.: Recent trends, opportunities and challenges of biodiesel in Malaysia: an overview. Renew. Sustain. Energy Rev. **14**(3), 938–954 (2010)
93. Lim, B.Y., et al.: A review of processing and machinery for Jatropha curcas L. fruits and seeds in biodiesel production: Harvesting, shelling, pretreatment and storage. Renew. Sustain. Energy Rev. **52**, 991–1002 (2015)
94. Gil, A.: Challenges on waste-to-energy for the valorization of industrial wastes: electricity, heat and cold, bioliquids and biofuels. Environ. Nanotechnol. Monitor. Manag. **17**, 100615 (2022)
95. Ajanovic, A.: Biofuels versus food production: does biofuels production increase food prices? Energy. **36**(4), 2070–2076 (2011)
96. Park, H.Y., et al.: Comparisons of combustion characteristics between bioliquid and heavy fuel oil combustion in a 0.7 MWth pilot furnace and a 75 MWe utility boiler. Energy. **192**, 116557 (2020)
97. Slade, R., Bauen, A.: Micro-algae cultivation for biofuels: cost, energy balance, environmental impacts and future prospects. Biomass Bioenergy. **53**, 29–38 (2013)
98. Maity, J.P., et al.: Microalgae for third generation biofuel production, mitigation of greenhouse gas emissions and wastewater treatment: present and future perspectives–A mini review. Energy. **78**, 104–113 (2014)
99. Dutta, K., Daverey, A., Lin, J.-G.: Evolution retrospective for alternative fuels: first to fourth generation. Renew. Energy. **69**, 114–122 (2014)
100. Joshi, G., et al.: Challenges and opportunities for the application of biofuel. Renew. Sustain. Energy Rev. **79**, 850–866 (2017)
101. Alsultan, A.G., et al.: A short review on catalyst, feedstock, modernised process, current state and challenges on biodiesel production. Catalysts. **11**(11), 1261 (2021)

102. Kenney, K.L., et al.: Understanding biomass feedstock variability. Biofuels. **4**(1), 111–127 (2013)
103. Goswami, L., et al.: A critical review on prospects of bio-refinery products from second and third generation biomasses. Chem. Eng. J. **448**, 137677 (2022)
104. Khan, S., et al.: Challenges and perspectives on innovative technologies for biofuel production and sustainable environmental management. Fuel. **325**, 124845 (2022)
105. Singh, J., Gu, S.: Commercialization potential of microalgae for biofuels production. Renew. Sustain. Energy Rev. **14**(9), 2596–2610 (2010)
106. Amerit, B., et al.: Commercialization of biofuel products: a systematic literature review. Renew. Energy Foc. **44**, 223–236 (2023)
107. Martin, M., et al.: Assessing the aggregated environmental benefits from by-product and utility synergies in the Swedish biofuel industry. Biofuels. (2020)
108. Ziolkowska, J.R.: Biofuels technologies: an overview of feedstocks, processes, and technologies. Biofuels for a More Sustainable Future, pp. 1–19 (2020)
109. Morone, P., Cottoni, L., Giudice, F.: Biofuels: technology, economics, and policy issues. In: Handbook of Biofuels Production, pp. 55–92. Elsevier (2023)
110. Ponte, S.: 'Roundtabling'sustainability: Lessons from the biofuel industry. Geoforum. **54**, 261–271 (2014)
111. Harnesk, D., Brogaard, S., Peck, P.: Regulating a global value chain with the European Union's sustainability criteria–experiences from the Swedish liquid transport biofuel sector. J. Clean. Prod. **153**, 580–591 (2017)
112. Lozano-García, D.F., et al.: GIS-based modeling of residual biomass availability for energy and production in Mexico. Renew. Sustain. Energy Rev. **120**, 109610 (2020)
113. Kumar, A., Sharma, S.: Potential non-edible oil resources as biodiesel feedstock: an Indian perspective. Renew Sustain. Energy Rev. **15**(4), 1791–1800 (2011)
114. Venkataramana, S.H., et al.: Niger seed oil-based biodiesel production using transesterification process: experimental investigation and optimization for higher biodiesel yield using box–behnken design and artificial intelligence tools. Appl. Sci. **12**(12), 5987 (2022)
115. Sahoo, P., Das, L.: Process optimization for biodiesel production from Jatropha, Karanja and Polanga oils. Fuel. **88**(9), 1588–1594 (2009)
116. Muthukumaran, C., et al.: Process optimization and kinetic modeling of biodiesel production using non-edible Madhuca indica oil. Fuel. **195**, 217–225 (2017)
117. Ogunkunle, O., Oniya, O.O., Adebayo, A.O.: Yield response of biodiesel production from heterogeneous and homogeneous catalysis of milk bush seed (Thevetia peruviana) oil. Energy Policy Res. **4**(1), 21–28 (2017)
118. Dubey, S.M., Gole, V.L., Gogate, P.R.: Cavitation assisted synthesis of fatty acid methyl esters from sustainable feedstock in presence of heterogeneous catalyst using two step process. Ultrason. Sonochem. **23**, 165–173 (2015)
119. Awolu, O.O., Layokun, S.K.: Optimization of two-step transesterification production of biodiesel from neem (Azadirachta indica) oil. Int. J. Energy Environ. Eng. **4**, 1–9 (2013)
120. Sai, B.A., et al.: Optimization of continuous biodiesel production from rubber seed oil (RSO) using calcined eggshells as heterogeneous catalyst. J. Environ. Chem. Eng. **8**(1), 103603 (2020)
121. Mathiarasi, R., Partha, N.: Transesterification of soap nut oil using novel catalyst. J. Saudi Chem. Soc. **21**(1), 11–17 (2017)

Chapter 2
Studies on Oil Extraction, Biodiesel Production Methods, and Engine Performance and Emission Characteristics

2.1 Introduction

This chapter provides a comprehensive literature review on oil extraction, biodiesel production methods, and their impact on engine performance and emission characteristics. It discusses various oil extraction techniques, their advantages, limitations, and criteria for selecting the most suitable method. The transformation of crude oil into biodiesel and the variables affecting their effectiveness are also explored. The transesterification process is analysed, focusing on key factors influencing biodiesel conversion and methods to control these factors. Various experimental methods for optimizing biodiesel production and engine performance are reviewed, comparing classical approaches with advanced statistical design techniques like Taguchi and Response Surface Methodology. The use of Artificial Intelligence tools to enhance biodiesel performance is also examined. The chapter provides a framework for experimentation, modelling, and optimization to improve both performance and emissions in diesel engines.

2.2 Framework Illustrating Selection Criteria for Biodiesel Production

Figure 2.1 illustrates the framework established for effective biodiesel production. Feedstock alone accounts for approximately 70–95% of biodiesel production cost; therefore, feedstock selection is of industrial and economic relevance. The choice of an appropriate oil extraction method that could maximize the oil extraction from feedstock is of industrial significance. Studying and selecting the best biodiesel conversion technologies from extracted oils maximizes the biodiesel yield, which

M. Patel G. C. et al., *Biofuel Production, Performance, and Emission Optimization*, Springer Optimization and Its Applications 215,
https://doi.org/10.1007/978-3-031-84806-3_2

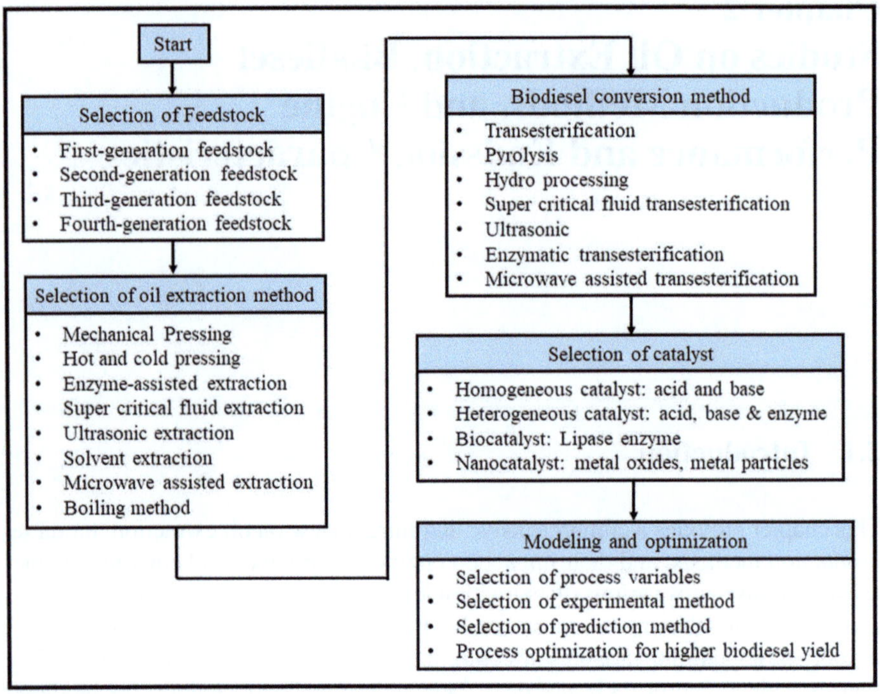

Fig. 2.1 Framework for effective biodiesel production

could lower the overall production cost. The conversion efficiencies are relevant to the appropriate choice of catalyst that influences the chemical reactions. Studying and analysing the process variables helps industry personnel to understand the process insights. The proper selection of experimental methods helps industry personnel analyse process insights with fewer experimental trials at a reduced cost. Furthermore, establishing a model that derives an input-output relationship helps industry personnel predict outputs. Finally, modelling and optimization focus on fine-tuning the biodiesel production process by selecting appropriate process variables and experimental and prediction methods to achieve higher biodiesel yields.

2.3 Introduction to the Oil Extraction Process

An essential stage in the production of biodiesel is the extraction of oil from both edible and inedible seeds. This can be accomplished by solvent extraction or mechanical pressing, and the resulting oil is a valuable feedstock for the process of turning the seeds into biodiesel, which is an eco-friendly and renewable substitute for traditional diesel fuel.

2.3.1 Methods of Oil Extraction Process

- **Mechanical Pressing**

  Mechanical pressing, executed through a device like an expeller press or screw press, involves applying pressure to crush and squeeze oil from seeds. It's a common method for extracting oils from both edible and non-edible seeds. The mechanical force exerted ruptures the cell walls of the seeds, facilitating the release of oil. This straightforward and efficient technique is favoured for its versatility, making it applicable to various oilseed types. However, the suitability of mechanical pressing depends on factors such as seed type, oil content, and the desired quality of the extracted oil [1–4].
- **Solvent Extraction**
  The solvent extraction process is employed for seeds with low oil content and when maximum yield is desired. Hexane, ether (diethyl and petroleum), and ethanol are typical solvents used for oil extraction. The solvent is then separated, leaving behind the extracted oil. Solvent extraction, such as the Soxhlet apparatus offers numerous advantages in recycling during the oil extraction process [5, 6].
- **Cold Pressing**
  Cold pressing is a method in which seeds undergo mechanical pressing at temperatures below 122 °F (50 °C). This process preserves the oil's natural flavour, colour, and nutritional value, making it a preferred technique for producing high-quality edible oils. By avoiding high temperatures, cold pressing minimizes the risk of heat-induced degradation, ensuring the final product retains its inherent qualities, making it suitable for culinary applications where the maintenance of oil's original characteristics is crucial [7–9].
- **Hot Pressing**
  Hot pressing is a method in which seeds undergo preheating before mechanical pressing, lowering oil viscosity for enhanced extraction efficiency. The application of heat facilitates the separation of oil from seeds, particularly beneficial for varieties with higher oil content. This process optimizes oil flow during pressing, resulting in improved yields. While effective, hot pressing should be carefully managed to prevent the degradation of oil quality due to excessive heat. This method is widely employed in industrial settings, offering an efficient means of extracting oils from seeds with elevated oil concentrations [10, 11].
- **Enzyme-Assisted Extraction**
  Enzyme-assisted extraction employs enzymes to break down cell walls, enhancing the efficient release of oil from seeds. This method is favoured for its ability to improve extraction yields by promoting the breakdown of complex structures within the seeds. Enzymes act as catalysts, accelerating the process and making oil extraction more effective. This environmentally friendly approach is particularly useful for enhancing the extraction of oils from various sources, contributing to increased efficiency in the production of edible oils, biofuels, and other applications [12–14].

- **Supercritical Fluid Extraction (SFE)**
  SFE employs supercritical fluids, such as carbon dioxide (CO_2), to extract oils. Operating in a supercritical state, CO_2 exhibits properties of both a liquid and a gas, making it an efficient solvent. This method is acclaimed for its ability to produce high-quality oils, as the supercritical CO_2 offers precise control over extraction parameters while minimizing thermal degradation. SFE is widely utilized in the food, pharmaceutical, and cosmetic industries, providing a clean and selective extraction process that preserves the integrity of bioactive compounds in oils, making it suitable for various applications [15–17].
- **Ultrasonic Extraction**
  Ultrasonic extraction utilizes high-frequency ultrasonic waves to disrupt cell structures, speeding the release of oil from seeds. The mechanical vibrations produced by ultrasonic waves induce cavitation, which breaks cell walls and improves oil extraction efficiency. This method is known for its effectiveness in significantly reducing extraction times compared to traditional techniques. The non-thermal and selective nature of ultrasonic extraction minimizes degradation, preserving the quality of the extracted oil. Widely used in the food, pharmaceutical, and cosmetic industries, ultrasonic extraction provides a fast and efficient process, making it a valuable technique for various applications that requires high-quality oil extraction [18–20].
- **Microwave-Assisted Extraction**
  Microwave-Assisted Extraction utilizes microwaves to heat seeds and rupture cell walls, facilitating the efficient release of oil. This method stands out for its energy efficiency and reduced extraction times compared to conventional processes. Microwaves penetrate the seeds, generating internal heat, and promoting the separation of oil from cellular structures. MAE's rapid and selective heating contributes to minimized degradation, resulting in oils with enhanced quality. Widely recognized for its eco-friendly attributes and time-saving advantages, microwave-assisted extraction is increasingly adopted in industries such as food, cosmetics, and pharmaceuticals for its effectiveness in obtaining high-quality oils [21–23].
- **Boiling Method**
  Seeds are boiled in water to release oil by initially reducing moisture followed by grinding the seeds to form a paste and extracting, scooping, and drying the oil subjected to heat. The boiling method is typically used for sesame seeds with low oil content [24, 25].

2.3.2 *Advantages and Limitations of Oil Extraction Process (Table 2.1)*

Table 2.1 Advantages and limitations of oil extraction methods

Method	Advantages	Disadvantages
Mechanical pressing method	• Mechanical pressing results in lower operating or processing costs • This method does not require heat or chemicals and reduces the risk of contamination • Mechanical oil extraction method produces high-quality oil with a low concentration of FFAs • This method does not use heat (unlike the boiling method) for the oil extraction process and therefore preserves heat-sensitive nutrients (vitamins and antioxidants) in the oil	• Large portion of oil is left in the cake or meal after extraction • Time-consuming and labour-intensive • Mechanical pressing can remove 45% of the oil and may require assistance from the solvent method for the residual oil present in the meal • High capital investment is needed for building essential infrastructure such as specialized equipment • Risk of contamination from equipment
Solvent extraction method	• The solvent extraction method ensures higher yield of oil conversion from seeds, resulting in an effective oil extraction process • This efficiency method is due to the lower proportion of residual oil left in the cake	• Oil conversion is more expensive than mechanical pressing due to higher solvent costs • High capital investment and longer duration for oil extraction consume more energy • This is not an environmentally friendly procedure because volatile organic compounds are emitted into the atmosphere during the extraction process • The atmospheric release of volatile organic compounds
Boiling method	• Cost-effective method of producing oil • Requires minimal equipment and energy compared to other methods	• Tedious and time-consuming (energy-sapping, drudgery prone, and inefficient) • Resulting in lower yield and quality • Boiling for a longer duration at higher temperatures results in the degradation of oil • Higher operating temperatures result in a higher risk, such as burns or injuries

2.3.3 *Criteria for Selection of Oil Extraction Process*

The selection of an oil extraction process is influenced by various factors that ensure optimal efficiency, yield, and quality. Key criteria for choosing an extraction method include:

- **Seed Type**
  Seed type is a critical criterion for selecting an oil extraction method. Variability in oil content and characteristics among different seeds dictates the extraction process suitable for each. Compatibility is key, as certain methods may be more effective for high-oil-content seeds, while others suit seeds with specific properties. Understanding the unique composition of each seed guides the choice of an extraction process, ensuring optimal efficiency and yield. Factors such as seed hardness, size, and the presence of protective coatings impact the method's success, highlighting the necessity of tailoring extraction techniques to the specific traits of each seed type [26].
- **Oil Yield**
  The desired oil yield is a pivotal criterion in choosing an oil extraction process. The efficiency of the selected method directly impacts the quantity of oil obtained from the raw material. High oil yield is often a priority for economic and practical reasons. Factors such as extraction method, seed type, and processing conditions influence overall efficiency. Striking a balance between optimal oil yield and resource utilization ensures the chosen extraction process aligns with production goals, whether for edible oils, biofuels, or industrial applications. The consideration of oil yield is essential in optimizing the economic viability of the extraction process [27].
- **End Product Quality**
  End product quality is a critical factor influencing the selection of an oil extraction process. The intended application, whether for food, biofuels, or industrial purposes, determines the necessary quality standards. For edible oils, stringent criteria for flavour, colour, and nutritional content apply. Biofuels demand high purity to meet fuel specifications, while industrial applications may prioritize specific chemical properties. Tailoring the extraction method to meet these quality requirements ensures that the produced oil aligns with the intended purpose, meeting market standards and consumer expectations. The consideration of end product quality is paramount in determining the success and viability of the chosen extraction process [28].
- **Economic Viability**
  Economic viability is a pivotal criterion in selecting an oil extraction process, encompassing factors such as initial investment, operational costs, and energy efficiency. The choice of extraction method should align with budget constraints while optimizing resource utilization. Equipment expenses, maintenance costs, and energy requirements influence the overall economic feasibility. Striking a

balance between efficiency and affordability ensures a sustainable and profitable extraction process. Evaluating the long-term financial implications, including potential returns on investment, is essential for making economically sound decisions in the oil extraction industry. The consideration of economic viability is integral to creating a sustainable and cost-effective production process [29].

- **Environmental Impact**
  Environmental impact is a critical consideration in oil extraction, with a growing emphasis on sustainable and eco-friendly practices. Minimizing ecological consequences involves choosing extraction methods that reduce habitat disruption, water usage, and carbon emissions. Employing renewable energy sources and minimizing chemical solvent usage are integral to environmentally conscious processes. The industry's commitment to reducing its carbon footprint and adopting green technologies reflects a broader acknowledgment of the importance of mitigating environmental impact. By prioritizing sustainability, the oil extraction sector contributes to a more eco-friendly and responsible approach, aligning with global efforts to preserve and protect the environment [30, 31].
- **Scale of Production**
  The scale of production is a pivotal factor shaping the selection of an oil extraction process. Whether operating on a small scale or an industrial level, the choice of method and equipment must align with production goals. Small-scale operations may prioritize simplicity and cost-effectiveness, while industrial-scale processes demand efficiency and high throughput. Tailoring the extraction approach to the scale of production ensures optimal resource utilization, cost-efficiency, and overall effectiveness. This consideration acknowledges the diverse needs of varying production scales, enabling the implementation of extraction methods that best suit the scope and capacity of the operation [32].
- **Energy Efficiency**
  Energy efficiency is a crucial factor in selecting an oil extraction process. The energy consumed during extraction impacts both the sustainability and cost-effectiveness of the operation. Methods that minimize energy consumption and utilize renewable energy sources contribute to a more sustainable and environmentally friendly approach. Evaluating the energy requirements, considering the availability of power sources and adopting energy-efficient technologies are integral to optimizing the overall efficiency of the extraction process. Striking a balance between energy efficiency, cost considerations, and environmental impact ensures a responsible and economically viable approach to oil extraction [33, 34].
- **Extraction Speed**
  Extraction speed is a crucial consideration in selecting an oil extraction method, particularly in response to production demands. The efficiency of the extraction process directly relates to how quickly oil can be obtained from the raw material. Industries with high production requirements prioritize methods that offer rapid extraction, ensuring timely and continuous output. Balancing speed with other factors, such as yield and quality, is essential to meet efficiency requirements.

Tailoring the extraction method to achieve the desired speed aligns with production goals, optimizing the overall effectiveness and responsiveness of the oil extraction process [35–37].

- **By-product Utilization**
  By-product utilization is a key criterion in oil extraction, contributing to a sustainable and resource-efficient process. Extraction methods that effectively repurpose by-products, like seed cakes or glycerol, minimize waste and enhance the overall ecological footprint. Seed cakes can be repurposed as animal feed or fertilizers, while glycerol, a by-product of biodiesel production, finds application in various industries. Adopting methods that prioritize the utilization of by-products aligns with environmentally conscious practices, promoting circular economies, and reducing the environmental impact of the oil extraction process. It reflects a commitment to maximizing the value derived from the raw materials, ensuring a more holistic and responsible production cycle [38, 39].
- **Quality Preservation**
  Quality preservation is paramount in selecting an oil extraction process, particularly for applications like edible oils. Minimizing heat exposure during extraction is essential to retain the oil's natural flavour, colour, and nutritional value. Methods that prioritize quality preservation, such as cold pressing or supercritical fluid extraction, ensure the final product meets stringent standards. By preventing thermal degradation, these processes maintain the integrity of the oil, meeting consumer expectations for high-quality and nutritious products. The consideration of quality preservation is instrumental in producing oils suitable for culinary use and other applications where the original characteristics are vital [40].
- **The Availability of Raw Materials**
  The availability of raw materials is a pivotal factor influencing the selection of an oil extraction method. Regions with abundant and accessible oilseed crops may prefer methods that are well-suited to local cultivation, promoting economic efficiency and reducing transportation costs. The choice of extraction process should align with the prevalent oilseed varieties, ensuring a sustainable and locally viable production cycle. Considering the availability of raw materials facilitates the optimization of the extraction process, supporting regional economies and contributing to the overall feasibility and success of oil production in a given geographical area [41, 42].

2.4 Biodiesel Conversion or Production Methods

The biodiesel production methods from the feedstocks are summarized with potential advantages and limitations targeting low-cost production (refer to Table 2.2). The biodiesel production methods are evaluated, highlighting their benefits and limitations. The transesterification process is recommended as potential biodiesel conversion.

Table 2.2 Advantages and limitations of biodiesel production methods from crude oil

Production Methods	Advantages	Limitations	References
Catalytic distillation: Vacuum and molecular	Ease of separation	The biodiesel conversion rate after post-treatment relies on catalyst recovery and the solvent used	[86]
Reactive distillation	Ease of separation of products Simple process for conversion of feedstock containing high FFA	Catalyst efficiency decides the conversion rate Higher energy requirements	[86]
Microwave technology	Faster reaction process Less heat loss	The catalyst must be removed to ensure complete biodiesel conversion	[86]
Dilution	Simple process	Incomplete combustion resulting in carbon deposition	[87]
Micro-emulsion	Simple process	High viscosity Less stable	[88]
Pyrolysis	Simple process Less emission	The high temperature required for the conversion process and the higher cost Generates carbon residue	[89]
Super fluid method	Higher conversion efficiency ensures faster reaction The conversion method does not require a catalyst	Installation cost is higher Requires higher energy as a result of higher pressure and temperature	[90]
Transesterification	Fuel properties are in good agreement with diesel Higher conversion rate The technique can be easily applied to an industrial scale	Low free fatty acids feedstocks Require higher water content Due to the neutralization and water washing of products Difficult reaction product separation	[91]

2.4.1 *Methods for Biodiesel Conversion from Crude Oil*

Biodiesel is typically generally derived from triglycerides found in plant oils or animal fats. Different catalysts (acidic, alkaline, and enzymatic) are available for biodiesel conversion [43]. The choice of catalyst to be used for biofuel conversion depends on the free fatty acids and the nature of the feedstocks [44]. The advantages and disadvantages of various catalysts are discussed in Table 2.3. Generally, homogeneous catalysts have an advantage over heterogeneous catalysts in terms of reaction rate. However, heterogeneous catalysts are preferred as they are recovered, allow for continuous operation, are not sensitive to free fatty acid, and can be reused

Table 2.3 Comparison of homogeneous and heterogeneous catalyst

Homogeneous		Heterogeneous	
Alkaline catalyst [81]	Acid catalyst [82]	Alkaline catalyst [83, 84]	Acid catalyst [85]
NaOH, KOH, CH_3ONa, CH_3OK	$H_2(SO)_4$, HCl	CaO, MgO, SrO, $Mg_6Al_2(OH)16CO_3.4(H_2O)$	ZrO_2, SnO_2, TiO_2, zeolites, sulphonic ion-exchange resins, $ZrO_2/Al_2O_3/WO_3$, SiO_2/ZrO_2, SO_2–4 /TiO_2/La_3+
Advantages			
• Abundant availability of catalyst • The reaction takes place at atmospheric conditions (pressure and temperature) • High reaction rate	• Resistance to FFA. • Simultaneous process of both transesterification and esterification	• Low-cost catalyst synthesis • Possibility of catalyst regeneration and reuse • Similar to homogeneous acid catalysts • Environmental safety	• Resistance to FFA • Simultaneous process of both transesterification and esterification • Less toxic • Less corrosion compared to a homogeneous acid catalyst • Reuse of catalyst
Disadvantages			
• Sensitivity to FFA and water • Initial purification of feedstock is required	• Undergoes high-temperature reaction • Require higher ratio of alcohol-to-oil • Long process duration • Promote corrosion of equipment and materials	• Undergoes high-temperature reaction • By-products formation • Slow reaction	• Slow process • Formation of undesired reactions • High catalyst cost

[45]. Using a homogeneous alkali catalyst results in a reaction rate 4000 times faster than a heterogeneous catalyst and is available at a low cost than an acid catalyst [46]. Enzymatic catalysts (bacterial, fungal, and yeast lipases) are an alternate choice in the transesterification process for biofuel conversion. However, there are significant disadvantages such as high production costs for enzymes, slow reaction rates, enzyme inhibition in alcohol, limited enzyme regeneration, and gradual deactivation. Based on the literature study, homogeneous catalysed transesterification is an effective method for producing biodiesel. The selection of transesterification parameters is crucial for converting feedstocks into biodiesel [47]. Therefore, it is important for the industry to identify the control variables that maximize biodiesel yield and minimize costs. Here are the methods for biodiesel conversion from various feedstocks (Fig. 2.2):

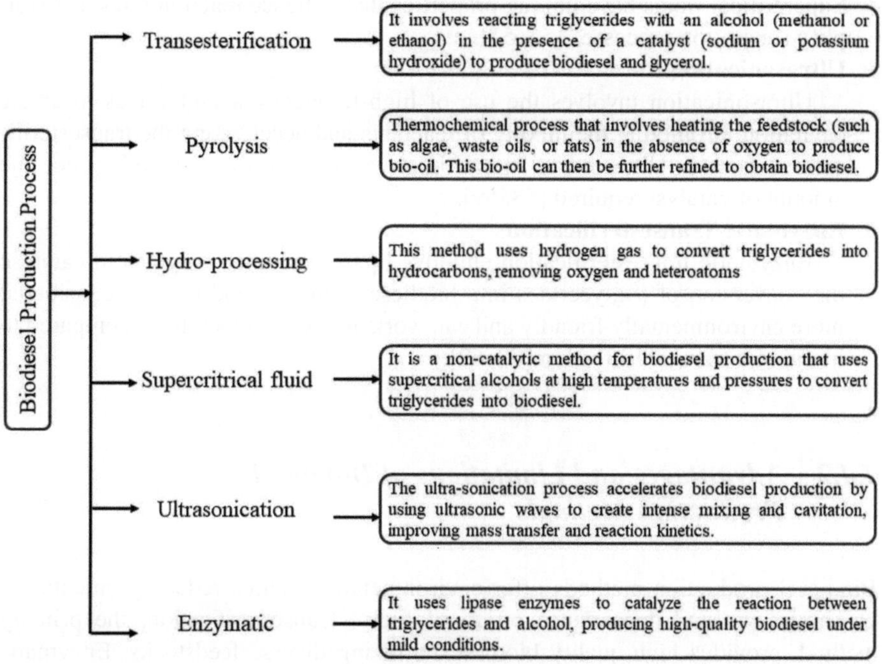

Fig. 2.2 Biodiesel conversion process from various feedstocks

1. **Transesterification**

   Transesterification is the most common method for converting triglycerides (fats and oils) into biodiesel. It involves reacting triglycerides with an alcohol (usually methanol or ethanol) in the presence of a catalyst (usually sodium or potassium hydroxide) to produce biodiesel and glycerol. The chemical reaction results in the separation of the fatty acids from the glycerol backbone, forming biodiesel (methyl or ethyl esters) and glycerine [48–50].
2. **Pyrolysis**

   Pyrolysis is a thermochemical process that involves heating the feedstock (such as algae, waste oils, or fats) in the absence of oxygen to produce bio-oil. This bio-oil can then be further refined to obtain biodiesel. Pyrolysis has the advantage of being able to process a variety of feedstocks, including waste materials, but it requires careful control of temperature and pressure [51–53].
3. **Hydroprocessing**

   Hydroprocessing involves treating triglycerides with hydrogen in the presence of a catalyst, typically nickel or palladium, to yield hydroprocessed biodiesel. This method is more commonly used for refining conventional diesel but can be adapted for biodiesel production. The process removes impurities and increases the stability of the biodiesel [54, 55].
4. **Supercritical Fluid Transesterification**

   This method involves the use of supercritical fluids, such as supercritical methanol or supercritical ethanol, to carry out the transesterification reaction.

Supercritical fluids have unique properties that enhance reaction rates and simplify separation processes [56, 57].

5. **Ultrasonication**

   Ultrasonication involves the use of high-frequency sound waves to create cavitation, promoting the mixing of reactants and accelerating the transesterification reaction. Ultrasonication can improve reaction kinetics and reduce the amount of catalyst required [58, 59].

6. **Enzymatic Transesterification**

   Enzymatic transesterification employs lipases or other enzymes to catalyse the conversion of triglycerides into biodiesel. This method is often considered more environmentally friendly and can work at lower temperatures compared to traditional transesterification [60–62].

2.4.2 *Advantages and Limitations of Biodiesel Production Methods*

Biodiesel production methods offer environmental benefits, reducing greenhouse gas emissions and dependence on fossil fuels. Transesterification, the primary method, provides high-quality biodiesel, utilizing diverse feedstocks. Enzymatic processes are environmentally friendly, while Hydroprocessing ensures improved stability. However, challenges include feedstock availability, competition with food production, and the energy-intensive nature of certain methods. Glycerol by-product disposal remains a concern. Overall, biodiesel production methods contribute to sustainable energy, but careful consideration of feedstock sources, efficiency, and environmental impact is essential for widespread adoption. Table 2.2 lists the advantages and limitations of biodiesel production methods from crude oil.

2.4.3 *Selection Criterion for Effective Biodiesel Production Method*

Choosing an optimal biodiesel production method demands a holistic assessment, weighing factors like feedstock availability, conversion efficiency, and environmental impact. Sustainability, cost-effectiveness, and adherence to quality standards are paramount. The selected method should efficiently convert feedstock into biodiesel, ensuring high yields while minimizing energy consumption and waste. Compatibility with existing infrastructure, scalability, and regulatory compliance further influence the decision. Achieving a balance between economic viability and ecological considerations is crucial in the pursuit of effective biodiesel production. Here are key factors to consider when choosing a biodiesel production method:

1. **Feedstock Availability and Suitability**

Assessing feedstock availability and sustainability is pivotal in biodiesel production. opt for abundant, renewable feedstocks that avoid competition with food production, ensuring ethical sourcing. The chosen method must accommodate variations in feedstock composition, displaying adaptability to diverse raw materials. A robust evaluation of feedstock accessibility and compatibility establishes the foundation for a biodiesel production process that aligns with environmental and ethical considerations while maximizing resource efficiency [63–65].

2. **Conversion Efficiency**
   Evaluating biodiesel conversion efficiency is paramount, encompassing factors such as biodiesel yield per unit of feedstock, reaction completeness, and the method's energy efficiency. A high conversion efficiency ensures optimal resource utilization, minimizing waste and maximizing biodiesel production. Assessing the overall effectiveness of the conversion process is essential for sustainable biodiesel production, promoting economic viability and reducing environmental impact. Efficient processes contribute to the viability of biodiesel as a renewable and cleaner alternative to traditional fossil fuels [66, 67].
3. **Reaction Kinetics**
   Delving into the reaction kinetics is pivotal when selecting a biodiesel production method. The speed of chemical reactions profoundly influences productivity and processing time. Methods with faster reaction rates enhance efficiency, allowing for higher biodiesel yields and shorter production cycles. A careful consideration of reaction kinetics ensures not only increased productivity but also streamlined processes, contributing to the economic viability and competitiveness of the chosen biodiesel production method [68, 69]. Fast reaction kinetics plays a crucial role in advancing the sustainability and practicality of biodiesel as an alternative, environmentally friendly energy source.
4. **Catalyst Selection**
   The selection of catalysts profoundly influences biodiesel production. Assessing the type of catalyst is crucial—homogeneous catalysts like sodium or potassium hydroxide, and heterogeneous catalysts such as solid catalysts, offer distinct advantages and drawbacks. Homogeneous catalysts often exhibit higher reaction efficiency, but separation can be challenging, impacting catalyst reuse. Meanwhile, heterogeneous catalysts facilitate easier separation and recycling but may have slower reaction rates. A careful consideration of catalyst properties is essential, striking a balance between efficiency, separation ease, and catalyst reusability to optimize the chosen biodiesel production method [70, 71].
5. **Purity of Biodiesel**
   Ensuring the purity of biodiesel is pivotal in the selection of production methods. The chosen approach must minimize impurities, adhering to stringent quality standards like ASTM. Maintaining high biodiesel purity is essential for compatibility with existing diesel engines, preventing engine issues and ensuring optimal performance. A method that prioritizes quality control not only aligns with industry standards but also promotes consumer confidence and wider acceptance of biodiesel as a reliable and efficient alternative to traditional fossil

fuels. Rigorous quality considerations are fundamental for the successful integration of biodiesel into existing fuel infrastructure [72–74].

6. **Glycerol Separation**

   The evaluation of glycerol separation holds paramount importance in biodiesel production. As a by-product, efficient separation directly impacts the economic viability of the process. Streamlining the removal of glycerol, a key element in transesterification reactions, enhances overall productivity and contributes to the cost-effectiveness of biodiesel manufacturing. Ensuring ease and efficiency in glycerol separation not only supports economic feasibility but also addresses environmental concerns by minimizing waste. A production method that effectively manages glycerol by-product reinforces the sustainability and competitiveness of biodiesel as a cleaner and renewable alternative to traditional fossil fuels [59, 75, 76].
7. **Environmental Impact**

   Thoroughly assessing the environmental impact is vital when selecting a biodiesel production method. Factors such as energy consumption, waste generation, and the use of hazardous chemicals play a crucial role. Methods that minimize these impacts are generally favoured for their sustainability. Choosing low-impact processes aligns with eco-friendly practices, reducing the carbon footprint and contributing to environmental conservation. A method with minimal environmental consequences not only complies with regulatory standards but also supports the broader goal of sustainable energy production, making biodiesel a more environmentally responsible choice in the transition away from conventional fossil fuels [77, 78].
8. **Scale-up Potential**

   Evaluating the scalability of a biodiesel production method is essential. While some methods may excel in small-scale or laboratory settings, their effectiveness may diminish when scaled up to industrial levels. Factors such as reaction kinetics, feedstock handling, and process control become critical at larger scales. Ensuring that the selected method can seamlessly transition from small to industrial production is vital for sustained efficiency and economic viability. A production approach with inherent scalability not only facilitates growth but also supports the integration of biodiesel as a substantial contributor to the broader energy landscape [79, 80].
9. **Cost Analysis**

   A comprehensive cost analysis is indispensable when selecting a biodiesel production method. Factors such as raw material costs, catalyst expenses, energy consumption, and operational outlays must be meticulously considered. Striking a balance between production costs and biodiesel yield is paramount for economic viability. Opting for a method that ensures cost-effectiveness contributes to the competitiveness of biodiesel in the energy market. A thorough economic evaluation not only guides decision-making but also positions biodiesel as a financially sound and sustainable alternative to conventional fuels, fostering its wider adoption in the pursuit of greener and more economically viable energy solutions [68].

2.5 Introduction to Transesterification Process with Advantages and Limitations

Transesterification is a pivotal chemical process at the heart of biodiesel production, converting triglycerides found in vegetable oils or animal fats into biodiesel and glycerol. This reaction involves the replacement of glycerol with an alcohol, typically methanol or ethanol, in the presence of a catalyst such as sodium or potassium hydroxide or solid nano-catalyst. The result is the formation of fatty acid methyl or ethyl esters, which constitute biodiesel, along with glycerol as a valuable by-product (when using homogeneous catalyst).

One of the notable advantages of transesterification is its ability to harness renewable resources, contributing to the reduction of reliance on non-renewable fossil fuels. Biodiesel, produced through this process, serves as a cleaner-burning alternative to conventional diesel, exhibiting similar combustion properties. Despite its prevalence and efficiency, challenges include feedstock variability, energy intensity, and the need for meticulous catalyst handling. As research and innovation continue, transesterification remains a fundamental and evolving technique in the pursuit of sustainable and environmentally friendly biodiesel production.

2.5.1 Identifying Factors Influencing the Biodiesel Conversion

1. **Methanol-to-Oil Molar Ratio**

   The optimal values for the M:O ratio range between 1:5 and 1:12 (except for 1:27.6 in [92] and 1:30 in [93]) for maximizing biodiesel yield from different feedstocks. The optimal conditions vary differ for the same feedstock and different feedstocks due to the following reasons:

   - Fatty acids compositions vary among different feedstocks.
   - Identical feedstocks sourced from different locations may have different levels of free fatty acids.
   - Different catalysts and proportions (weight/volume) are used under varying reaction conditions (reaction time, temperature, stirring speed).
   - Various experimental and optimization are employed.

   In general, higher fatty acid content requires a large amount quantity of alcohol for the reaction process [94]. However, the alcohol ratio depends on the catalyst and type of oil. While higher biodiesel conversion can be achieved with a higher M:O ratio and reduced reaction time, it can lead to drawbacks such as increased cost, no significant improvement beyond a critical M:O ratio, methanol excess methanol nullifying catalytic activity, and challenges in separation and purification [95].

2. **Catalyst Loading**

   The optimal values corresponding to different catalysts vary between 0.5 and 8 wt.% (except 9 wt.% for [96] and 11 wt.% for [97]) towards higher biodiesel conversion for different feedstocks. The catalyst type and proportion in the transesterification process affect the chemical reaction that causes hydrolysis and soap formation. An increased proportion of catalyst concentration leads to higher biodiesel conversion due to an increased proportion of triglyceride and ester). However, a higher catalyst concentration can result in agglomeration, creating more inactive sites that limit biofuel conversion (i.e. the slurry becomes viscous and emulsifier) limitation [98]. For example, the catalyst concentration varied between 2 and 10% wt./vol. during experimentation resulting in 97% biodiesel conversion at 8% wt./vol [97]. Similar observations are made for different feedstocks when analysing the catalyst loading in the transesterification process.

3. **Reaction Temperature**

   The authors reported that the optimal temperature conditions varied in the ranges of 40–65 °C (except for 75 °C in [99] and 80 °C in [100]) for biodiesel conversion from different feedstocks (refer to Table 2.4). Beyond the critical temperature of 80 °C, the alcohol may boil during the transesterification reaction [101]. If the transesterification experiments are carried out at room temperature, more time is required for biodiesel conversion from crude oil [102]. An increase in temperature increases the reaction rate and decreases reaction time. In Table 2.4, for the fixed reaction temperature of 65 °C, the reaction time is 420 min with a lipase catalyst [103] and 51.42 min with Bio-alkaline catalyst [104]. The reaction temperature varies for different feedstocks, and the parametric conditions employed for experiments (catalyst type and its concentration, reaction time, alcohol) are probably due to the different properties of feedstocks. Therefore, the optimal reaction temperature that maximizes biodiesel conversion from crude oil is of industrial relevance.

4. **Reaction Time**

   The time required for biodiesel conversion with chemical reactions for different feedstocks varied from 15 min to 240 min (except 300 min in [105], 360 min in and 420 min in [99]) is presented in Table 2.4. The reaction time differs for different catalysts because of catalyst size and proportion. It was observed that KOH [106] and NaOH catalyst [107], and nano-catalysts require lesser reaction time (Ni-ZnO in [97], Mn-ZnO in [97]). Increased chemical reaction time increased fatty acid conversions [108]. Conversely, glycerol formation from crude oil reduces biodiesel conversions beyond the critical value [109]. Many researchers reported that transesterification undergoes reversible reactions (triglycerides conversion to other compounds) during biodiesel conversion resulting in decreased biodiesel yield conversion. Although biodiesel yield increases with reaction time, prolonged exposure to methanol and catalyst reduces the biodiesel yield. Therefore, optimal reaction time for feedstocks is of practical importance.

5. **Stirring Speed**

   The operating range of the stirring speed is found to vary in the range between 150 and 600 rpm (refer to Table 2.4). In order to start and improve the chemical

Table 2.4 Experimental and optimization approaches for different feedstocks

Feedstock	Catalyst	EM	CC	M:O	Reaction condition	Yield (%)	OM	References
Castor oil	Si-MMT- pH-SO_3H	OFAT	5 wt.%	1:12	60 °C/300 min	89.8	OFAT	[105]
	Ni-ZnO nano-catalyst	RSM	11 wt.%	1:8	55 °C/60 min	95.2	RSM	[97]
Mahua oil	Mn-ZnO nano-catalyst	OFAT	8 wt.%	1:7	50 °C/50 min	97.0	OFAT	[111]
	NaOH	RSM	0.9 wt.%	1:8	55 °C/120 min	91.3	RSM	[128]
	NaOH	RSM	0.3 wt.%	1:4.25	53.9 °C/148.8 min	96.3	GA	[128]
Cottonseed oil	Ti/SiO_2 nano-catalyst	OFAT	5 wt.%	1:30	65 °C/240 min	98	OFAT	[93]
	CH_3NaO	RSM	0.97 wt.%	1:6	63.8 °C/797 rpm	98.3	RSM	[129]
Jatropha oil	SrO/CaO-H	BBD	7 mmol/g	1:27.6	65 °C/89.8 min	99.7	RSM	[92]
	KOH	NR	1.5 wt.%	1:12	15 min	99	ANN	[130]
Castor-karanja oil	H_2SO4/KOH	Taguchi	6 wt%	NR	50 °C/120 min/600 rpm	99.2	Taguchi	[115]
Karanja oil	CH_3KO	BBD	1 wt.%	1:9	60 °C/35 min/600 rpm	91.6	RSM	[126]
Rubber oil	Sodium metasilicate/CaO	OFAT	9 wt.%	1:9	60 °C/40 min/200 rpm	97	OFAT	[96]
	CaO	CCD	5 wt.%	1:9	65 °C/240 min	97.8	ANN	[123]
	Al_2O_3/CaO	CCD	3 wt.%	1:9	65 °C/240 min	98.9	ANN	[131]
	CaO	CCD	4 wt.%	1:12	65 °C/180 min	99.7	ANN	[132]
Date seed oil	CaO	OFAT	5 wt.%	1:12	65 °C/90 min	93.5	OFAT	[112]
	Fe_2O_3 coated mercaptoacetic acid	RSM	1.5 wt.%	NR	55 °C/47 min	91.4	RSM	[133]
Madhuca indica oil	Copper@dolomite	OFAT	3 wt.%	1:15	75 °C/360 min	96.1	OFAT	[99]
	Na-CaO	Taguchi	5 wt.%	1:9	60 °C/60 min	81.4	Taguchi	[116]
Jojoba oil	NaOH	CCD	1.5 wt.%	1:16	60 °C/15 min	71.2	DFA	[107]
	Lipase [103]	OFAT	5 wt.%	1:5	65 °C/420 min	95.1	OFAT	[103]
Linseed oil	CaO	Taguchi	7 wt.%	1:8	60 °C/240 min	93	GRA	[117]
Flaxseed oil	KOH	CCD	0.51 wt.%	1:5.9	59 °C/33 min	98	RSM	[106]
	Bio-alkaline	CCD	2.7 wt.%	1:11	65 °C/51.42 min	96.5	RSM	[104]

(continued)

Table 2.4 (continued)

Feedstock	Catalyst	EM	CC	M:O	Reaction condition	Yield (%)	OM	References
Moringa oil	MgO nano-catalyst	Taguchi	1 wt.%	1:12	45 °C/240 min	93.7	Taguchi	[118]
	CuO-CaO	CCD	0.85 vol.%	1:1	60 °C/70.2 min	95.3	RSM	[121]
Neem oil	Fe, Ni, and Cu-doped ZnO	OFAT	2 wt.%	1:5	60 °C/80 min/150 rpm	80–95	OFAT	[113]
	KoH	BBD	1.02 wt.%	1:7.1	41 °C/98 min	96.4	RSM	[124]
Soap nut	MgO-KOH	OFAT	3 wt.%	1:3	80 °C/120 min/300 rpm	94	OFAT	[100]
Pongamia oil	KOH	BBD	1.43 wt.%	1:11.06	56.6 °C/81.43 min	98.4	RSM	[125]
Tobacco seed oil	KOH	CCD	1.15 wt.%	1:5.95	56.6 °C/77.6 min	90.2	GWO	[122]
	KOH	NR	1.1 wt.%	1:6	80 min	95	NR	[134]
	Tungstophosphoric acid	NR	4.4 wt.%	1:6	65 °C/240 min	87.5	NR	[135]

OFAT = one-factor-at-a-time, RSM = response surface methodology, GA = genetic algorithm, NR = not reported, DFA = desirability function approach, GRA = grey relational analysis, BBD = box-Behnken design, GWO = grey wolf optimization reaction between the catalyst and oil, stirring speed (to control catalyst agglomeration) is necessary. No significant changes were observed experimentally in biodiesel conversion with the stirring speed [110]. Therefore, many researchers fixed the operating values of the stirring rate during transesterification reactions. However, determining the optimal stirring time and speed is essential for maximum biodiesel conversion.

2.5.2 *Identify the Methods Employed for Controlling Factors Affecting Biodiesel Conversion*

Biodiesel production involves a chemical process called transesterification, where triglycerides (found in feedstock oils) react with an alcohol (usually methanol or ethanol) in the presence of a catalyst to form methyl or ethyl esters (biodiesel) and glycerol. Controlling factors such as reaction time, temperature, stir speed, alcohol-to-oil molar ratio, catalyst type and loading is crucial for optimizing the efficiency and yield of the biodiesel production process. The control and optimization of these factors are essential for enhancing biodiesel yield and quality. Additionally, optimizing these factors requires a balanced approach that considers the cost, efficiency, and environmental impact of the biodiesel production process. Various methods have been applied to control and optimize factors affecting biodiesel conversion, focusing on experimental, prediction, and optimization techniques.

2.6 Literature Review on Experimental Methods for Analysis and Optimization on Biodiesel Production, Performance and Emission Characteristics

Table 2.4 describes the experimental (to establish an input-output relationship) and optimization (to maximize biodiesel yield) approaches for different feedstocks. The methods used are classical engineering experimental or OFAT approach, DOE and AI techniques. Selection of the appropriate method results in enhanced performance (biodiesel yield) and reduced processing costs. Each method poses its potential advantages and limitations, which are discussed below.

2.6.1 Classical Engineering Experimental Approach or One-Factor-At-A-Time (OFAT)

OFAT involves varying one variable at a time and keeping other factors fixed at middle or different levels. Experiments are performed by varying factors such as M:O ratio, catalyst concentration, reaction temperature, and reaction time and analysing the performance of biodiesel yield for feedstocks such as castor oil [105], mahua oil [111], cottonseed oil [93], rubber oil [96], date seed oil [112], madhuca indica oil [99], Jojoba oil [103], neem oil [113], and soap nut [100]. A summary of observations made by OFAT approaches is discussed below:

1. Experiments increase with variables and levels, resulting in energy waste (equipment, labour, and material) and costs.
2. The OFAT approach needs to provide detailed insight into the interaction effects of process variables on the analysed response.
3. Practical guidelines for experimentation may not help industry personnel predict the outputs without investigation.
4. The optimal conditions determined using the OFAT approach result in local solutions. The optimal solutions are derived based on the experiments performed.

2.6.2 Statistical Design of Experimental Approach: Taguchi, Response Surface Methodology

DOE allows researchers to examine the effects of process variables, estimate the significance of parameters on outputs, and derive empirical relationships (relating input-output) with fewer experimental trials [114]. DOE provides detailed insight into the impact of input variables on a physical system's response. In 1964, Johnston introduced statistical methods applied for conducting experiments at a foundry. The popular DOE methods currently found in most literature are Full Factorial Design (FFD) and Fractional factorial designs (BBD, CCD, Taguchi). All methods are established to correlate inputs and outputs. The Taguchi method was applied to maximize the performance of biodiesel yield for the following feedstocks: Castor-Karanja oil [115], Madhuca indica oil [116], Linseed oil [117], and Moringa oil [118]. Although the Taguchi technique maximizes biodiesel yield cost-effectively with transesterification variables, it has the following drawbacks:

1. Taguchi matrix (experimental orthogonal array) is limited to testing the effect of all factors' interaction under investigation [119].
2. The Taguchi method limits monitoring and optimizes offline performance [120].
3. The Taguchi method is generally applied for the initial step of process development resulting in a one-time improvement technique that leads to local or sub-optimal solutions.
4. Taguchi's method failed to derive a valuable empirical equation for prediction.

5. The Taguchi method relies on other methods (such as GRA) for optimizing multiple outputs [117].

RSM methods (BBD and CCD) overcome most of the shortcomings of the Taguchi method and therefore applied them to optimize high-yield biodiesel (refer to Table 2.4). The CCD method is used to obtain high-yield biodiesel for the feedstocks, namely Jojoba oil [107], Flaxseed oil [106], Moringa oil [121], Tobacco seed oil [122], and Rubber oil [123]. BBD method was applied to maximize the biodiesel yield for the feedstocks such as Neem oil [124], Pongamia oil [125], Karanja oil [126], and Jatropha oil [92]. Although CCD and BBD data use the same mathematical model and are analysed by RSM, they differ in experimental conditions. CCD contains factorial experiments focused on the extreme corner point, i.e. factor levels, whereas BBD experiments are concentrated at the middle level of a factor [127]. Therefore, the selection of CCD (when the extreme combinations are essential to be analysed) and BBD (when the optimal combinations lie at the centre level of experiments and testing conditions at corner combinations are physically impossible) is decided by the investigator against the problem to be analysed. The following major observations were made:

1. The RSM technique needs to provide a detailed insight into a process.
2. RSM provide information on the curvature effect (linear or non-linear) on the measured.
3. RSM performs better in predicting with fewer independent variables (3–5). An increased number of experimental runs leads to a decrease in prediction accuracy to the increased number of variables.
4. RSM derives the empirical relationships for individual outputs and therefore cannot capture detailed insights into process mechanics and physics.
5. RSM is not an efficient technique for solving multiple response optimization, as the objective functions are non-linear and multi-modal.

2.6.3 Evaluation of Biodiesel Performance and Emission Characteristics in Diesel Engines

Diesel engines will be used to test the biodiesel produced from various feedstocks for its practicality. Table 2.5 presents comprehensive information about the performance and emission characteristics of biodiesel for different feedstocks, including waste oil and seed oil. Biodiesel fuelled to diesel engines resulted in approximately similar (slightly higher BSFC and lower BTE) performances close to diesel fuel. However, biodiesel's emission performances (excluding NOx) are comparatively better than diesel fuel. Therefore, diesel engines fuelled with biodiesel have potential environmental benefits (reduced emissions) and cost benefits (local availability of feedstock materials at low cost and improved energy security for the near future) benefits [136]. To date biofuel production has been produced from edible sources

Table 2.5 Summary results of performance and emission characteristics of biodiesel derived from different feed stocks

Feedstock	Experimental and optimization method	Input variables studied	Performance parameters	Emission parameters	Remarks (compared to neat diesel)	References
Jatropha biodiesel	Taguchi	CBD, SL, CR, IT, PZCD, A:F, PZHRC, DZCD DZHRC, C-o-R	BTE	NS	The CR, DZHRC, and DZCD are the major contributing factor for BTE	[137]
Karanja biodiesel	FFD and DFA	CR, BT, BP	BTE, BSFC	CO, HC, NOx	Optimized condition resulting in BTE, BSFC, CO, HC, and NOx is 33.65%, 0.2718 $kg/kW^{-1} h^{-1}$, 0.109%, 158, and 938 ppm	[142]
Jatropha biodiesel	CCD and NSGA-II	BT, BP, CR	BSFC, BTE, Pmax	CO, HC, NOx, SO	RSM-derived regression equations help to determine global solutions, viz., NSGA-II	[144]
PDSO	OFAT	BP	BSFC, BTE	CO, HC, NOx	↑ in BSFC, ↓ in BTE, ↓ in NOx, CO, HC	[145]
Pongamia biodiesel	RCCD and DFA	BT, EL, IP, IT	BTE	HC, NOx	↓ in IT, and ↑ in IP result in ↓ of HC and NOx, and ↑ in BTE	[146]
Polanga biodiesel	Taguchi	BT, IT, IP	BTE	HC, NOx, SO	↑ in BTE and NOx, ↓ in HC, SO	[147]
Palm biodiesel	Taguchi and fuzzy	EL, BT	BTE, BSFC, BSEC	HC, NOx	↑ in EL ↓ in BSEC. NOx has negligible effect with EL. ↑ in HC for all BT	[148]
Palm biodiesel	OFAT	BP	BSFC, BTE	CO, CO_2, NOx, HC	The performance and emission characteristics of biodiesel is better than diesel	[149]
Soy biodiesel	OFAT and GA, and PSO	EGR, PT, PR, IT, IP	BSFC	NOx, HC, CO, PM	PSO-GA produced better performance and emission characteristics than GA	[150]

Karanja biodiesel	Taguchi and GRA	EL, CR, BT	BTE	NOx	Grey-Taguchi method optimized condition produced better results	[151]
Karanja biodiesel	Taguchi and GRA	EL, BT, CR	BP, BSFC, BTE	CO, HC, NOx, CO_2, O_2	Higher values of EL, BT, and CR resulting in better response values	[152]
Karanja biodiesel	FFD and DFA	CR, BT, BP	BTE, BSFC	CO, HC, NOx	Optimized condition resulting in BTE, BSFC, CO, HC, and NOx is 33.65%, 0.2718 $kg/kW^{-1}\ h^{-1}$, 0.109%, 158, and 938 ppm	[142]
Karanja biodiesel	Taguchi	IA, CR, BT, EL	BTE, BSFC, EGT	CO_2, NOx, CO, HC	↓ in BTE by 2%, ↑ in 3% by BSFC and EGT, ↑ of 0.86% for CO_2, ↓ in 0.029% for CO, ↓ in 8% for HC and ↓ in 12 PPM for NOx.	[153]
Castor biodiesel	OFAT, and NSGA-II and TOPSIS	ES, BT	BSFC, BP	PM, CO, NOx, CO_2	NSGA-II produced many solutions TOPSIS select the best solution	[154]
Mahua biodiesel	Taguchi and GRA	EL, BT	BSFC, BTE, EGT	HC, CO, O_2, NOx, SO, CO_2	Taguchi-GRA method optimizes all the six objective functions	[155]
Corn biodiesel	Taguchi	EL, BT, EGR, ES	BSFC	NOx, SO, CO, CO_2	Applying Taguchi method save 70% of time with regard to time and cost	[156]
Cotton seed biodiesel	Taguchi and GRA	EL, ES, BT	BP, BSFC	CO, HC, NOx	B20 + higher EL and lower ES resulting in better response values	[157]
Soybean biodiesel	Taguchi and GRA	BT, WC, Al NPs	BSFC, BSEC	CO, HC, NOx, SO	Taguchi-GRA determined optimized conditions resulting in better output values	[158]
Soybean biodiesel	Taguchi and GRA	BT, WC, ZnO NPs	BSFC, BTE	CO, HC, NOx, SO	B20 + higher WC and ZnO NPs resulting in optimized conditions	158]
Jatropha biodiesel	FFD and GA	IT, EL, ES	BSFC, BTE, Pmax	CO, HC, NOx, SO	GA determined optimized conditions resulting in better response values	[159]

(continued)

Table 2.5 (continued)

Jatropha biodiesel	CCD and DFA	EL, CR, IP	BP, BTE	HC, NOx	↓ in HC, ↑ in NOx was observed	[160]
Jatropha biodiesel	Taguchi	CBD, SL, CR, IT, PZCD, A:F, PZHRC, DZCD DZHRC, C-o-R	BTE	NS	The CR, DZHRC, and DZCD are the major contributing factor for BTE	[137]
Jatropha biodiesel	CCD and NSGA-II	BT, BP, CR	BSFC, BTE, Pmax	CO, HC, NOx, SO	RSM-derived regression equations help to determine global solutions, viz., NSGA-II	[144]
Biodiesel derived from waste feed stocks						
Waste fish oil biodiesel	BBD and DFA	BT, WC, PS	BP, BSFC, BTE	CO, HC, NOx, CO_2	DFA is applied to optimize both performance and emission characteristics	[138]
Waste cooking oil biodiesel	CCRD and ABC	EL, ES, BT	BP, BSFC	CO, HC, NOx	ABC determine the single optimal condition satisfying five response values	[161]
Waste cooking oil biodiesel	Taguchi and ANN, GRA	BT, ES, EL	BP, BSFC	CO, HC, NOx, CO_2	B100, with higher ES and EL, resulting in best response values	[162]
Biodiesel derived from different feed stocks						
Biodiesel	RSM and ANN	IP, BT, EL	BSFC, BTE, EGT	NOx, HC, CO, SO	ANN predict the performance and emission parameters accurately	[163]
Biodiesel	OFAT and GOA	ET, BT, EL	BSFC, BTE	CO, HC, NOx, CO_2	GOA determined the optimized condition for better response values	[164]
Biodiesel	CCD and GA	BT, ES, EL	BP, BSFC	CO, NOx, SO	GA showed their effectiveness in optimizing the input parameters	[165]

Mixed oil biodiesel feedstock						
Rubber—Palm biodiesel	OFAT	BT, ES	EGT, BSFC, BTE, BP	CO, NOx, CO_2	Performance and emission characteristics of biofuel are close to diesel fuel	[136]
Soybean + sunflower biodiesel	OFAT	EL, BT	BTE, BSFC, BSFEC, EGT	CO_2, NOx, CO, HC, O_2	↑ in 30–70% of BT resulting in ↓ in CO, CO_2, EGT, BSEC and HC, ↑ in NOx, and BSFC	[140]
Jatropha – algae biodiesel	OFAT	EL	BSFC, BTE	CO, HC, NOx, O_2	↑ in EL resulting in ↓ in BSFC, CO, O_2. ↑ in EL resulting in ↑ in BTE, HC, NOx	[140]
Waste cooking—Soybean biodiesel	CCD and DFA	EL, BT, EGR	BSFC, BTE	CO, HC, NOx, SO	Better coefficient of determination and desirability value obtained for each response	[141]
Rubber—Palm biodiesel	CCD and DFA	EL, ES, BT	BSFC, BTE, BP, EGT	CO, HC, SO, NOx, CO_2, O_2	DFA determined optimized condition for all outputs with a desirability value equal to 0.96	[166]

ABC = artificial bee colony, A:F = air-to-fuel ratio, Al NPs = aluminium nanoparticles, BP = brake power, BT = blend type, BSFC = brake-specific fuel consumption, BTE = brake thermal efficiency, CBD = cylinder bore diameter, CO = carbon monoxide, CR = compression ratio, C-o-R = cut-off ratio, DFA = desirability function approach, DZHRC = diffusive zone heat release constant, DZCD = diffusive zone combustion duration, EGR = exhaust gas recirculation, EGT = exhaust gas temperature, EL = engine load, ET = ethanol, FFD = full factorial design, GOA = grasshopper optimization algorithm, HC = hydrocarbon, IA = injection angle, IP = injection pressure, IT = injection timing, NOx = nitrogen oxide, NS = not studied, NSGA-II = non-dominated sorting genetic algorithm-II, PM = particulate matter, Pmax = peak cylinder pressure, PR = pilot ratio, PS = percent of surfactant, PT = pilot timing, PDSO = pithecellobium dulce seed oil, PZHRC = premixed zone heat release constant, PZCD = premixed zone combustion duration, RCCD = rotary central composite design, SL = stroke length, SO = smoke opacity, WC = water concentration, ZnO NPs = zinc oxide nanoparticles (palm, soybean, sunflower, cotton seed, corn, etc.) and non-edible feedstock sources (Karanja, Pongamia, castor, Mahua, Jatropha). However, these sources are not recommendable due to land limitations (food vs fuel war), price fluctuations, energy policies, social movements, and industry growth limited to using edible source-based biodiesel for a longer run [137]. Therefore, waste feedstocks derived biodiesel is tested in diesel engines [138]. Although different feedstocks are derived for biodiesel production, they do not cater to the high demand needs for industrial-scale production. Consequently, mixing edible and inedible oils, as well as waste cooking and fish oils, offers an alternative option to meet the feedstock requirements for the manufacturing of biodiesel. Examples of oil mixtures include Rubber-palm [136], Jatropha-Algae [139], Soybean-sunflower [140], and waste cooking-soybean [141]. Biodiesel generated from mixed oils has demonstrated its potential for use as a substitute fuel source for diesel engines. Rotatable Central Composite Design (RCCD or CCRD) is applied with all factors set at five levels. The prediction accuracy of the rotatable CCD output is stable around the design centre. The choice between rotatable and non-rotatable designs depends on geometric design space constraints. Unless practical conditions dictate otherwise, rotatable designs may necessary. Many studies use non-rotatable CCD or BBD designs to statistically analyse biodiesel the performance in diesel engines. FFD statistically analyses the performance and emissions of Karanja and Jatropha biodiesel in diesel engines [142]. However, FFD requires more experimental trials compared to fractional factorial (CCD and BBD) experimental designs. RSM applied to experimental designs (CCD, BBD, FFD) derive empirical relationships (i.e. regression equations relating input and output), which include full quadratic terms (linear, square, and interaction). The derived regression equations were used for predictions and as objective functions for optimization tasks. Taguchi experimental designs were applied to estimate only the main factor effects, neglecting important interaction factors and square terms. Leaving out all square and few interaction terms results in imprecise input-output relationships that reduce prediction accuracies [143]. The regression equations created from the experimental designs (CCD, BBD, and FFD) were used as a goal to achieve multi-objective optimization of performance and emission features [144]. Experiments were conducted (using various methods such as OFAT, CCD, BBD, FFD, and Taguchi with a set of parameters including blend type (percentage of diesel and biodiesel), engine speed and load, injection timing and pressure. The performances (BSFC, BSEC, BTE, EGT) and emission characteristics (CO, CO_2, HC, NOx) of diesel engines were analysed. Soft computing approaches such as ANN, GA, PSO, GOA, Fuzzy, and NSGA-II were utilized to predict and optimize the response functions, leading to improved response values.

Table 2.5 highlights the various experimental and optimization methods used to study the performance and emission characteristics of different biodiesel feedstocks. Optimization studies are conducted for each feedstock due to differences in fuel characteristics. Engine performance parameters (BTE, BSFC, BP, and EGT) and emissions (CO, HC, NOx, SO, CO_2, and O_2) are influenced by key variables such as CR, BSFC, BP, EL, IT, and IP. Full factorial and fractional factorial experimental designs (Taguchi, CCD, CCRD, BBD) are applied to reduce the

Table 2.6 CEEA matrix for four factors and three levels (−1, 0, +1)

	Input parameters				Responses					
Exp. no	A	B	C	D	R_1	–	–	–	–	R_6
Exp. 1	−1	0	0	0						
Exp. 2	0	0	0	+1						
–	0	0	+1	0						
–	0	−1	0	0						
–	0	+1	0	0						
–	0	0	−1	0						
–	+1	0	0	0						
–	0	0	0	−1						
Exp. 9	0	0	0	0						

Exp. 1, ...Exp. 9 are experimental runs. R1......R6: Output variables
A, B, C, D: Input variables; and Levels: low (−1), Middle (0), and high (1)

experimental plan, analyse parameters, and derive predictive equations. The performance of biodiesel compared to diesel shows improvements in emission characteristics (except for a slight increase in NOx) and a negligible percent decrease in BTE and BSFC. Optimized conditions obtained from GRA and TOPSIS are based on experimental trials used for model development, resulting in local solutions. Global solutions for emission and performance characteristics are determined using artificial intelligence and machine learning algorithms such as GA, PSO, NSGA-II, ANN, GOA, ABC, and others.

2.6.3.1 Classical Engineering Experimental Approach (CEEA)

To collect data on a single variable during each experimental run, the CEEA uses the OFAT concept while keeping all other factors constant. The aforementioned procedure is iterated until all other parameters under experimental investigation are tested. The CEEA steps are as follows:

1. After conducting laboratory experiments and reviewing the relevant literature, the range of factors to be investigated is determined.
2. Experiments are conducted by varying individual factors set at different levels while keeping the rest of the variables constant.

The experimental run for the traditional or CEEA is calculated using Eq. (2.1). The *n* and *k* term represent factors and levels. Table 2.6 displays the experimental matrix of the four variables and their coded three levels (low, medium, high: −1, 0, and 1). The term A, B, C, and D represent the input variables (transesterification variables: *methanol, sodium hydroxide, and reaction time;* engine variables*: speed, load,* and additive parameters: *diethyl ether, nanoparticles, and ethanol*) and R1 - R6 indicates the output variables (Responses: yield, NOx, CO, HC, BSFC, BP).

$$\text{Experimental run} = n(k-1)+1 \tag{2.1}$$

2.6.3.2 Advantages of CEEA

The CEEA method gathers data on the direct impacts of process factors. The significant influence of individual parameters on the responses examined by this approach allows for a preliminary screening of factors from a large number to a small number that do not significantly influence responses.

2.6.3.3 Limitations of CEEA

Comparing OFAT to experimental design reveals a limitations [167]:

1. It requires more experiments to estimate factor effects with the same precision.
2. It is unable to capture complete information on interaction effects, leading to analysis and conclusions that do not reveal the entire process.
3. Since it disregards the ideal situation or optimal solution, generalization is not possible.

In the early 1920s, Ronald Aylmer Fisher proposed statistical estimating techniques and experimental design theory [168]. Fisher introduced the concept of randomization into the design of experiments (DOE), where factors are simultaneously varied. The term "DOE" refers to a systematic approach to collecting experimental data, and applying regression analysis to interpret the data in order to derive conclusions about the defined mathematical input-output relationships of a physical systems [169]. The following are the most commonly used DOE techniques:

1. Factorial DOE
2. RSM-based DOE
   (a) Full Factorial Design (FFD)
   (b) Fractional Factorial Design (CCD and BBD)

2.6.3.4 Factorial DOE

Factorial experimental designs allow users to analyse the primary effect of process variables (average outcomes of individual factors over all treatments) and the combined effect (interaction). With additional variables to consider, complete factorial experimental designs require more experiments. However, the factor levels—two-level, three-level, and multi-level determine the number of trials to be undertaken. Two-level FFD are typically used to quickly determine the importance of primary and interaction effects among the many factors evaluated. Although two-level complete factorial studies can provide a thorough understanding of the main and individual influences on the measured response, they cannot recognize the curvature effect on the response function, if any. The slope and curvature effect on the response function when a factor changes levels depend on its three levels. The factorial points are recorded in the cube for two-level factorial experiments with three factors, as shown in Fig. 2.2a.

2.6.3.5 Response Surface Methodology

The response function curvature effect (if there is one) can be identified by setting the factors to at least three levels under investigation. The RSM was introduced in 1951 by Box and Wilson. RSM models and analyses industrial processes using statistical and mathematical approaches. The goal is to optimize the response (output) function by studying how different process variables affect measured responses. Regression analysis using experimental data helps users build mathematical models for input variables.

In the process of developing a regression equation using RSM, it is usual practice to make use of three essential terms. These terms are the main effect (linear), the square effect (quadratic), and the interaction term (combined). First-order models containing only the most critical effect terms are represented by Eq. (2.2) [169]:

$$Y_m = \beta_0 + \beta_1 x_{m1} + \beta_2 x_{m2} + \ldots\ldots\ldots\ldots, + \beta_k x_{mk} + \varepsilon_m \tag{2.2}$$

The first-order model, including the main and interaction variables, is depicted in Eq. (2.3).

$$Y_m = \beta_0 + \beta_1 x_{m1} + \beta_2 x_{m2} + \ldots, + \beta_k x_{mk} + \beta_{12} x_{m1} x_{m2} + .., \beta_{m-1,m} x_{mk-1} x_{mk} + \varepsilon_m \tag{2.3}$$

In the regression equation, the second-order polynomial comprises linear, square, and interaction terms, as illustrated in Eq. (2.4).

$$\begin{aligned} Y_m = \beta_0 + \beta_1 x_{m1} + \beta_2 x_{m2} + \ldots\ldots, + \beta_k x_{mk} + \beta_{11} x_{m1}^2 + \ldots.\ \beta_{kk} x_{mk}^2 + \beta_{12} x_{m1} x_{m2} \\ + \ldots., \beta_{m-1,m} x_{mk-1} x_{mk} + \varepsilon_m \end{aligned} \tag{2.4}$$

Equation (2.4) is rewritten in the standard form as shown in Eq. (2.5).

$$\underbrace{Y_m}_{\text{Response variable}} = \underbrace{\beta_0}_{\text{Regressor coefficient}} + \underbrace{\sum_{i=1}^{k} \beta_i x_{mi}}_{\text{Main effect terms}} \underbrace{\sum_{i=1}^{k} \beta_{ii} x_{mi}^2}_{\text{Quadratic effect terms}} + \underbrace{\Sigma \sum_{i<j} \beta_{ij} x_{mi} x_{mj}}_{\text{Interaction effect terms}} + \varepsilon_m \tag{2.5}$$

The term $m = 1, 2, \ldots, N$ (observations), where the mth observation corresponds to the factor depicted as. RSM uses the least-squares method to determine the regression equation coefficients. The mathematical formulation to establish the non-linear regression model and the procedure developed for experimental matrices of DOE is presented in the literature [169]. A non-linear regression model is developed for three levels for four factors, including all terms (linear, quadratic, and interaction) represented in general form as shown in Eq. (2.6).

Table 2.7 Coefficients, standard error, *T*-statistics, and *P*-values

Term	Model terms	Coefficient	SE Coef	T	P
Constant		--	--	--	--
X_1	Linear Terms	--	--	--	--
X_2		--	--	--	--
X_3		--	--	--	--
X_4		--	--	--	--
X_1^2	Square Terms	--	--	--	--
X_2^2		--	--	--	--
X_3^2		--	--	--	--
X_4^2		--	--	--	--
$X_1 X_2$	Interaction Terms	--	--	--	--
$X_1 X_3$		--	--	--	--
$X_1 X_4$		--	--	--	--
$X_2 X_3$		--	--	--	--
$X_2 X_4$		--	--	--	--
$X_3 X_4$		--	--	--	--

X_1 , X_2 ,X_3 , andX_4 are the factors represented in coded form

The term $m = 1,2,\ldots, N$ (observations), where the *m*th observation corresponds to the factor$x_i$depicted asx_{mi}. RSM uses the least-squares method to determine the regression equation coefficients. The mathematical formulation to establish the non-linear regression model and the procedure developed for experimental matrices of DOE is presented in the literature [169]. A non-linear regression model is developed for three levels for four factors including all terms (linear, quadratic, and interaction) represented in general form as shown in Eq. (2.6).

$$Y = \underbrace{\beta_0 + \beta_1 X_1 + \beta_2 X_2 + \beta_3 X_3 + \beta_4 X_4}_{\text{Linear terms with associated coefficients}} + \underbrace{\beta_5 X_1^2 + \beta_6 X_2^2 + \beta_7 X_3^2 + \beta_8 X_4^2}_{\text{Square terms with associated coefficients}} + \underbrace{\beta_9 X_1 X_2 + \beta_{10} X_1 X_3 + \beta_{11} X_1 X_4 + \beta_{12} X_2 X_3 + \beta_{13} X_2 X_4 + \beta_{14} X_3 X_4}_{\text{Interaction terms with associated coefficients}} \tag{2.6}$$

The statistical significance of the regression coefficients and the applicability of the derived models are evaluated once the regression model has been developed. The Design Expert software platform examines the factors' significance based on statistical computations (coefficients, standard error, *F*-statistics, and *p*-values) on different responses. Table 2.7 depicts the statistical analysis for four factors.

The following is a description of the terms included in Table 2.7: Coef term depicts coefficients used in Eq. (2.6), standard error coefficient (SE Coef) determines the forecasting precision. Note that the coefficients are more accurate when

Table 2.8 ANOVA for four factors and three levels

Source	DF	Sequenced SS	Adjusted SS	Adjusted MS	*F*	*P*
Regression	–	–	–	–	–	–
Linear (X_1, X_2, X_3, andX_4)	–	–	–	–	–	–
Square (X_1^2, X_2^2, X_3^2 and X_4^2)	–	–	–	–	–	–
Interaction (X_1X_2, X_1X_3, X_1X_4, X_2X_3, X_2X_4, X_3X_4)	–	–	–	–	–	–
Residual error	–	–	–	–	–	–
Lack of fit	–	–	–	–	–	–
Pure error	–	–	–	–	–	–
Total	–	–	–	–	–	–

DF degrees of freedom

the standard error is small. The ratio of corresponding values of Coef concerning the SE Coef determines the *T*-values. The independent variable's *T*-value can evaluate the response's parameter significance. *P*-value refers to the minimal threshold (95%) at which the importance of each process variable and their combined effect is evaluated. Furthermore, the *P*-value provides a quantitative measure of the significant impact of process variables under experimentation. A term in the regression model is considered significant if its underlying *P*-value is less than 0.05 at the 95% confidence level.

Analysis of variance (ANOVA) tests are used to estimate the coefficient of determination (R^2) and assess the robustness of the developed regression models statistically (Table 2.8). DF depicts the contributory terms that estimate the prediction error. Sequential SS measures the degree of variability introduced by each term in the data by summing their squares. The sum of squares estimated after removing non-significant factors from the model is called the Adjusted SS. MS (mean square) is calculated by dividing the sum of squares by the DF. In contrast, the adjusted MS (Adjusted MS) is evaluated after excluding the noncontributory terms from the regression equation. The F-value is the ratio of adjusted MS error to residual error, which is used to test the hypothesis.

Following the development of the regression model, the statistical appropriateness of the model is validated for practical applicability, as stated above. Experiments using parameter values varying between operating levels (high and low values) utilized for model development can be used to test the model's practical significance. Experimental conditions are generated randomly, and the outcomes are compared with the response values anticipated or predicted by the model to derive significant conclusions about the considered model.

The curvature effect can only be determined when the process variables are studied at least at three levels in the experimental design. These three-level designs include CCD, BBD, and Doehlert (DM) designs, which are predominantly second-order symmetrical response surface designs used for modelling three-level factorial designs. Interestingly, these symmetrical response surface designs differ in terms of

Table 2.9 Comparison of efficiency of response surface design models [169]

Factors	Number of coefficients (*P*)	Experimental run			Efficiency (*P*/ experimental run)		
		CCD	BBD	DM	CCD	BBD	DM
2	6	9	–	7	0.67	–	0.86
3	10	15	13	13	0.67	0.77	0.77
4	15	25	25	21	0.60	0.60	0.71
–	–	–	–	–	–	–	–
–	–	–	–	–	–	–	–
–	–	–	–	–	–	–	–
8	45	273	113	73	0.16	0.40	0.62

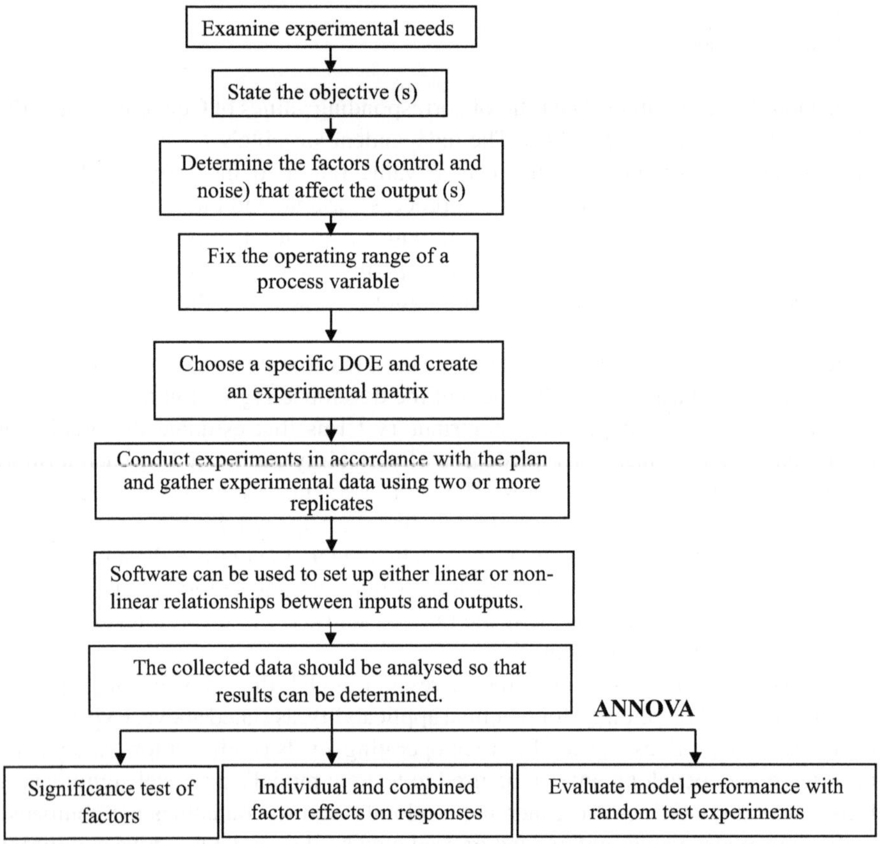

Fig. 2.3 RSM Steps

regarding the number of parameter levels, experimental points, experimental runs and blocks. Table 2.9 shows a comparison of the efficiency of response surface design models. The coefficients (*P*) in the full quadratic model for a given factor (*n*) can be estimated using Eq. (2.7).

Table 2.10 Comparison of CCDs [170]

Designs	Variable levels	Rotatable/ non-rotatable	Points outside ±1	Accuracy estimation
CCC	5	Rotatable	Y	Good over entire design space
CCI	5	Rotatable	N	Fair over central subset of design space
CCF	3	Non-rotatable	N	Fair over entire design space

Y yes, *N* no

$$P = \frac{(n+2)(n+1)}{2} \tag{2.7}$$

The flowchart shown in Fig. 2.3 illustrates the methodology adopted for modelling the biodiesel production and engine performance and exhaust emission process using RSM.

2.6.3.6 Central Composite Design

The response surface design model CCD, commonly referred to as the Box-Wilson design, is used to create complete quadratic models. There are three types of central composite designs (inscribed (CCI), circumscribe (CCC), and face-centred (CCF)). Specific designs (for instance, stars and factorial points) are selected based on the number of factors and necessary design elements. Table 2.10 summarizes a few of the critical qualities.

Either non-rotatable or rotatable designs are chosen depending on the geometrical nature of the real-world restrictions present in the design region. Rotatable designs require five levels of each factor, while non-rotatable methods require three. The CCD can be rotated if the generated models' output prediction accuracy is constant over the whole design space. Rotatable design should be used if it makes sense from a practical perspective [169]. Hence, a non-rotatable face-centred composite design is used for experimentation and analysis in the present work. CCD includes all combinations of two-level FFD when the parameters are set at their respective coded high (+1) and low (−1) levels and composed of the fractional part of eight-star points (allows estimation of curvature) together with three replicates at the centre point combination (coded value, 0). Thus, the face-centred CCD consists of 27 experimental runs for four independent variables, as shown in Table 2.11. The graphical representation of the three-level CCD for three factors is shown in Fig. 2.4b.

2.6.3.7 Box-Behnken Design

BBD is a fully or nearly rotatable second-order response surface design. The variables in BBD should be set at their three levels. CCD plan more experiments than BBD for the same input variables. The design avoids extreme situations that could

Table 2.11 Face-centred CCD matrix

	Input variable				Responses					
Exp. no	V_1	V_2	V_3	V_4	R_1	R_2	..	...	..	R_6
1	−1	−1	−1	−1						
2	0	0	0	0						
3	−1	0	0	0						
4	+1	−1	−1	−1						
5	−1	+1	−1	−1						
6	+1	0	0	0						
7	0	−1	0	0						
8	+1	+1	−1	−1						
9	−1	−1	+1	−1						
10	0	+1	0	0						
11	0	0	0	0						
12	−1	+1	+1	+1						
13	+1	+1	+1	+1						
14	0	0	0	+1						
15	0	0	0	0						
16	−1	+1	−1	+1						
17	0	0	+1	0						
18	−1	+1	+1	−1						
19	−1	−1	+1	+1						
20	−1	+1	+1	−1						
21	0	0	0	−1						
22	−1	−1	−1	+1						
23	+1	−1	−1	+1						
24	+1	−1	+1	−1						
25	+1	+1	−1	+1						
26	0	0	−1	0						
27	+1	−1	+1	+1						

V_1, V_2, V_3, and V_4 represent the variable 1, 2, 3, and 4

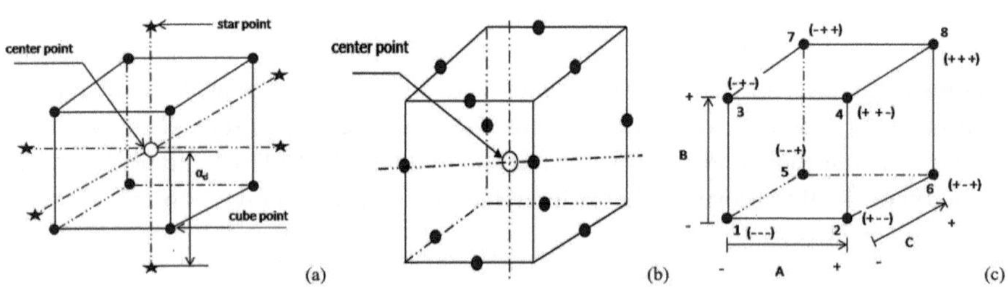

Fig. 2.4 Graphical representations showing designs for three factors: (**a**) two-level FFD, (**b**) three-level CCD, and (**c**) three-level BBD

Table 2.12 BBD matrix

	Input variable				Responses					
Exp. no	V_1	V_2	V_3	V_4	R_1	R_2	..	...	..	R_6
1	0	−1	0	−1						
2	0	+1	0	−1						
3	0	+1	−1	0						
4	−1	0	−1	0						
5	0	0	0	0						
6	0	0	0	0						
7	+1	0	0	+1						
8	−1	0	0	+1						
9	0	+1	0	+1						
10	0	0	0	0						
11	0	−1	−1	0						
12	0	0	+1	+1						
13	+1	+1	0	0						
14	+1	0	−1	0						
15	+1	0	0	−1						
16	−1	0	0	−1						
17	+1	0	+1	0						
18	0	−1	0	+1						
19	0	0	+1	−1						
20	−1	0	+1	0						
21	0	0	−1	+1						
22	−1	−1	0	0						
23	−1	+1	0	0						
24	0	0	−1	−1						
25	+1	−1	0	0						
26	0	+1	+1	0						
27	0	−1	+1	0						

lead to incorrect results by not varying factorial combinations at high or low levels. BBD is unsuitable for scenarios, especially when severe cube corner responses are needed. BBD consists of 27 experimental runs for four input variables (Table 2.12). Figure 2.4c displays the BBD graphically for three parameters.

2.6.3.8 Advantages of RSM Over Adaptive One Factor At-A-Time Method

The advantages of response surface methodology are as follows:

1. RSM allows the investigator to estimate full quadratic factor effects and a regression equation for a process. Conversely, OFAT requires more experiments to explain the system behaviour.
2. RSM estimates the impact of independent variable interactions on the dependent response.
3. RSM establishes mathematical input-output correlations using realistic experimental data, enabling accurate predictions and process optimization.
4. RSM provides comprehensive process insight.

2.6.3.9 Limitations of Response Surface Methodologies

1. RSM is not ideal for optimizing production processes due to the need for multiple tests, which can be costly for producers [170].
2. For optimizing non-linear systems and multi-modal objectives, mathematical iterative, heuristic, and metaheuristic search techniques outperform RSM.
3. RSM enables researchers to examine and formulate regression equations for each response. This approach estimates many responses simultaneously, which may lose output interdependency.

2.6.3.10 Artificial Intelligence Tools

Artificial intelligence (AI) is a computational tool used to simulate human intelligence (i.e. mimic the human mind, which is later programmed) by leveraging computers and machines for problem-solving and decision-making tasks. AI tools model and analyse the complex input-output relationship to understand the process. AI techniques overcome the disadvantages of hard computing to produce near-optimal solutions faster. Recently, optimization problems with single and multiple objective functions have been solved using metaheuristic methods. AI tools were applied for biodiesel research to evaluate the performance and emission characteristics. Fuzzy logic, ANNs, and machine learning optimization algorithms were the popular methods applied. ANNs were used for predicting of outcomes based on defined inputs. Fuzzy logic and algorithms (such as NSGA-II, GA, GOA, GWO, PSO, JAYA, and TLBO) can be applied to conduct optimization tasks (refer to Table 2.5). Figure 2.5 illustrates the classification of various optimization algorithms used in computational research. Metaheuristics are high-level problem-independent algorithmic frameworks that provide a set of guidelines or strategies to develop heuristic optimization algorithms. They are not directly inspired by natural or human phenomena but are rather general tactics designed to escape local optima and perform a robust

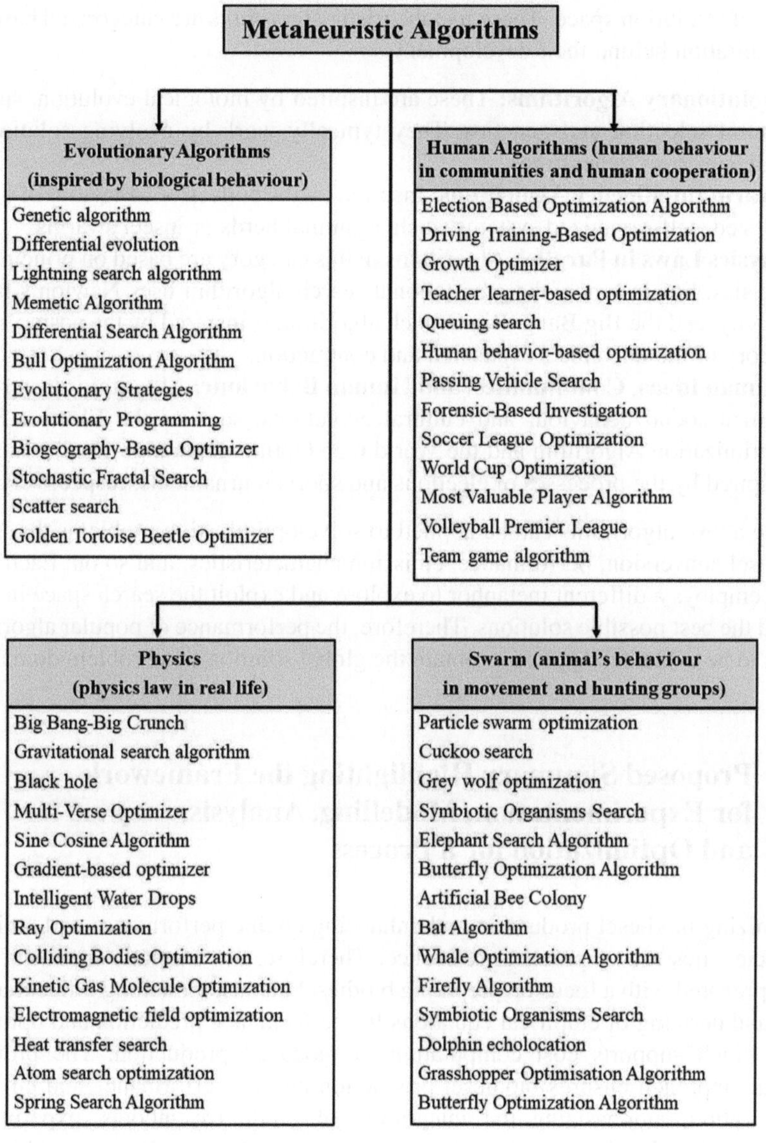

Fig. 2.5 Metaheuristic algorithms for optimization

search of a solution space. These metaheuristic algorithms are categorized based on the inspiration behind their development.

1. **Evolutionary Algorithms:** These are inspired by biological evolution, such as natural selection and genetics. They typically work by evolving solutions to problems over generations.
2. **Swarm Intelligence Algorithms:** Inspired by the collective behaviour of decentralized, self-organized systems, such as animal herds or insect swarms.
3. **Physics Laws in Parallel:** Algorithms in this category are based on principles of physics. For instance, the gravitational search algorithm uses Newton's law of gravity, and the Big Bang–Big Crunch algorithm is inspired by the cosmological theory of the universe's expansion and contraction.
4. **Human Ideas, Communities, and Human Behaviour:** Algorithms inspired by human social behaviour and cultural evolutions, such as the Election-Based Optimization Algorithm and the World Cup Optimization algorithm, which are inspired by the processes of elections and sports tournaments, respectively.

The above algorithms can be applied to solve optimization problems that affect biodiesel conversion, performance, emission characteristics, and so on. Each algorithm employs a different metaphor to explore and exploit the search space in order to find the best possible solutions. Therefore, the performance of popular algorithms needs to be evaluated applied to obtain the global solution to a problem domain.

2.7 Proposed Summary Highlighting the Framework for Experimentation, Modelling, Analysis, and Optimization for a Process

Optimizing biodiesel production and enhancing engine performance and emission characteristics are of practical relevance. Therefore, a systematic framework has been prepared with a focus on preparing biodiesel blends, modelling influential factors, and deriving of empirical equations for performance prediction and optimization, which supports cost computation of biodiesel production. The proposed holistic approach ensures biodiesel production and its performance and emission optimization, emphasizing the integration of predictive models, experimental design, and multi-objective optimization to enhance the overall process efficiency (refer to Fig. 2.6).

The following steps outline the process of converting seed oil into biodiesel and running it in a diesel engine:

1. Conversion of GGG seed oil into biodiesel via the transesterification process.
2. Identification of factors influencing biodiesel conversion through laboratory experiments and literature reviews.
3. Selection of an experimental design to study the effects of process variables on biodiesel conversion.

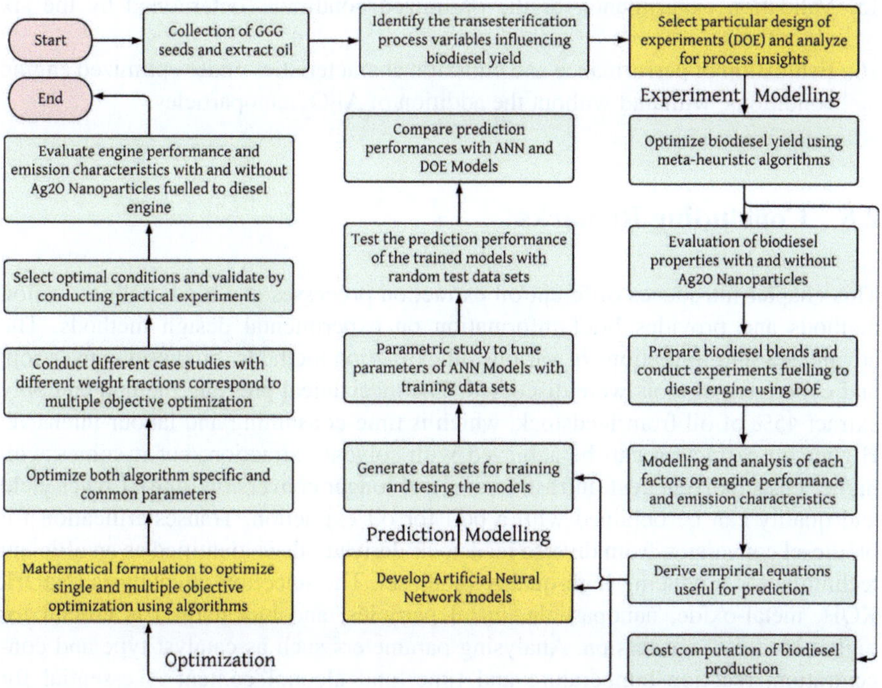

Fig. 2.6 A framework for biodiesel production, performance optimization, and emission reduction through the integration of experimental, predictive, and optimization mode

4. Execution of experiments according to the chosen design and model validation using tests such as the coefficient of determination, prediction, and optimization.
5. Optimization of biodiesel yield using six different metaheuristic algorithms.
6. Evaluation of fuel properties for both crude oil and optimized biodiesel, with and without Ag_2O nanoparticles.
7. Computation of cost analysis for the production of 1 L of biodiesel.
8. Testing biodiesel to assess its practical application in diesel engines.
9. Experimentation with biodiesel in diesel engines, examining a variety of engine parameters that influence emissions and performance characteristics.
10. Modelling and analysis of factors that affect engine performance and emissions.
11. Model validation for each performance and emission characteristic, utilizing tests such as the coefficient of determination, prediction, and optimization.
12. Development of an ANN (Artificial Neural Network) model for predicting control variables and emission and performance characteristics in both forward and reverse directions.
13. Mathematical formulation for optimization involving both single and multiple objective functions.
14. Optimization of engine variables using six different metaheuristic algorithms.

15. Validation experiments for the optimized conditions determined by the six algorithms.
16. Evaluation of performance and emission characteristics under optimized engine conditions, with and without the addition of Al_2O_3 nanoparticles.

2.8 Concluding Remarks

This chapter introduces different oil extraction processes and biodiesel conversion methods and provides brief information on experimental design methods. The advantages and limitations of various oil extraction methods, biodiesel conversion, and experimental tools were discussed. The mechanical pressing method can only extract 45% of oil from feedstock, which is time-consuming and labour-intensive. Higher oil extraction can be achieved with solvent extraction, but it comes with higher costs (solvent cost, infrastructure, and longer conversion time). Lower yield and quality can be obtained with a boil for oil extraction. Transesterification for biodiesel conversion from diverse feedstock-derived oils is identified as an efficient technique for producing high-quality biodiesel. The selection of catalysts (NaOH, KOH, metal-oxide, nanoparticle, metal particles, and biocatalyst) is crucial for higher biodiesel conversion. Analysing parameters such as catalyst type and concentration, reaction temperature and time, and alcohol content is essential for achieving higher yield. Engine and fuel-related parameter like compression ratio, brake-specific fuel consumption, brake power, energy loss, ignition timing, and injection pressure are important for improving performance metrics such as brake thermal efficiency, brake-specific fuel consumption, brake power, and exhaust gas temperature, as well as emission characteristics like CO, HC, NOx, SO, CO_2, and O_2. Biodiesel performance compared to diesel improves in terms of emission characteristics (except for a slight increase in NOx) with a negligible decrease in brake thermal efficiency and brake-specific fuel consumption. The chapter also covers statistical methods for testing the appropriateness of full factorial and fractional factorial experimental design models (Taguchi, CCD, CCRD, and BBD) to reduce the experimental plan, analyse parameters, and derive predictive equations. Optimized conditions derived from GRA and TOPSIS are based on experimental trials used for model development, resulting in local solutions. AI tools are computationally efficient in modelling complex input-output relationships and optimizing for both single-objective and multi-objective functions, which are thoroughly discussed. Global solutions for emission and performance characteristics are determined using artificial intelligence and machine learning algorithms such as GA, PSO, NSGA-II, ANN, GOA, ABC, and others.

References

1. Subroto, E., et al.: Mechanical extraction of oil from Jatropha curcas L. kernel: effect of processing parameters. Ind. Crop Prod. **63**, 303–310 (2015)
2. Kasote, D.M., Badhe, Y.S., Hegde, M.V.: Effect of mechanical press oil extraction processing on quality of linseed oil. Ind. Crop Prod. **42**, 10–13 (2013)
3. Savoire, R., Lanoisellé, J.-L., Vorobiev, E.: Mechanical continuous oil expression from oilseeds: a review. Food Bioproc. Tech. **6**, 1–16 (2013)
4. Ajith, B., et al.: Analysis and optimisation of transesterification parameters for high-yield Garcinia Gummi-Gutta biodiesel using RSM and TLBO. Aust. J. Mech. Eng. **20**(5), 1463–1478 (2022)
5. Fornasari, C.H., et al.: Efficiency of the use of solvents in vegetable oil extraction at oleaginous crops. Renew. Sustain. Energy Rev. **80**, 121–124 (2017)
6. Kumar, S.J., et al.: Green solvents and technologies for oil extraction from oilseeds. Chem. Cent. J. **11**, 1–7 (2017)
7. Cakaloglu, B., Ozyurt, V.H., Otles, S.: Cold press in oil extraction. A review. Ukrainian Food J. **7, Issue 4**, 640–654 (2018)
8. Saykova, I., et al.: Extraction of phenolic and flavonoid compounds from solid wastes of grape seed oil production by cold pressing. J. Chem. Technol. Metallurgy. **53**(2) (2018)
9. Khoddami, A., Man, Y.B.C., Roberts, T.H.: Physico-chemical properties and fatty acid profile of seed oils from pomegranate (Punica granatum L.) extracted by cold pressing. Eur. J. Lipid Sci. Technol. **116**(5), 553–562 (2014)
10. Siger, A., Józefiak, M., Górnaś, P.: Cold-pressed and hot-pressed rapeseed oil: the effects of roasting and seed moisture on the antioxidant activity, canolol, and tocopherol level. Acta Sci. Pol. Technol. Aliment. **16**(1), 69–81 (2017)
11. Ghazani, S.M., García-Llatas, G., Marangoni, A.G.: Micronutrient content of cold-pressed, hot-pressed, solvent extracted and RBD canola oil: implications for nutrition and quality. Eur. J. Lipid Sci. Technol. **116**(4), 380–387 (2014)
12. Nadar, S.S., Rao, P., Rathod, V.K.: Enzyme assisted extraction of biomolecules as an approach to novel extraction technology: a review. Food Res. Int. **108**, 309–330 (2018)
13. Liu, J.-J., et al.: Enzyme-assisted extraction processing from oilseeds: principle, processing and application. Innov. Food Sci. Emerg. Technol. **35**, 184–193 (2016)
14. Yusoff, M.M., Gordon, M.H., Niranjan, K.: Aqueous enzyme assisted oil extraction from oilseeds and emulsion de-emulsifying methods: a review. Trends Food Sci. Technol. **41**(1), 60–82 (2015)
15. Ahangari, H., et al.: Supercritical fluid extraction of seed oils–a short review of current trends. Trends Food Sci. Technol. **111**, 249–260 (2021)
16. Yousefi, M., et al.: Supercritical fluid extraction of essential oils. TrAC Trends Anal. Chem. **118**, 182–193 (2019)
17. Akanda, M.J.H., et al.: Applications of supercritical fluid extraction (SFE) of palm oil and oil from natural sources. Molecules. **17**(2), 1764–1794 (2012)
18. Thilakarathna, R., et al.: A review on application of ultrasound and ultrasound assisted technology for seed oil extraction. J. Food Sci. Technol. **60**(4), 1222–1236 (2023)
19. Junaid, P.M., et al.: Advances in seed oil extraction using ultrasound assisted technology: a comprehensive review. J. Food Process Eng. **46**(6), e14192 (2023)
20. Mushtaq, A., et al.: Advances in green processing of seed oils using ultrasound-assisted extraction: a review. J. Food Process. Preserv. **44**(10), e14740 (2020)
21. Zamanhuri, N.A., Abd Rahman, N., Bakar, N.F.A.: Effect of microwave power and extraction time on crude palm oil quality using microwave-assisted extraction process. Int. J. Renew. Energy Dev. **10**(3), 495 (2021)
22. Gomez, L., Tiwari, B., Garcia-Vaquero, M.: Emerging extraction techniques: microwave-assisted extraction. In: Sustainable Seaweed Technologies, pp. 207–224. Elsevier (2020)

23. Keskin Çavdar, H., et al.: Optimisation of microwave-assisted extraction of pomegranate (Punica granatum L.) seed oil and evaluation of its physicochemical and bioactive properties. Food Technol. Biotechnol. **55**(1), 86–94 (2017)
24. Tulashie, S.K., et al.: Effects of roasting and boiling on the yield, quality and oxidative stability of extracted soya bean oil. Thermal Sci. Eng. **1**(2) (2018)
25. Tongnuanchan, P., Benjakul, S.: Essential oils: extraction, bioactivities, and their uses for food preservation. J. Food Sci. **79**(7), R1231–R1249 (2014)
26. Veljković, V.B., Kostić, M.D., Stamenković, O.S.: Camelina seed harvesting, storing, pretreating, and processing to recover oil: a review. Ind. Crop Prod. **178**, 114539 (2022)
27. Halim, R., Danquah, M.K., Webley, P.A.: Extraction of oil from microalgae for biodiesel production: a review. Biotechnol. Adv. **30**(3), 709–732 (2012)
28. Das, P., et al.: A comprehensive review on the factors affecting thermochemical conversion efficiency of algal biomass to energy. Sci. Total Environ. **766**, 144213 (2021)
29. Markevičius, A., et al.: Trends and sustainability criteria of the production and use of liquid biofuels. Renew. Sustain. Energy Rev. **14**(9), 3226–3231 (2010)
30. Johnston, J.E., Lim, E., Roh, H.: Impact of upstream oil extraction and environmental public health: a review of the evidence. Sci. Total Environ. **657**, 187–199 (2019)
31. Mahfoud, F., et al.: Defatting and defatted peanuts: a critical review on methods of oil extraction and consideration of solid matrix as a by-product or intended target. Processes. **11**(8), 2512 (2023)
32. Pragya, N., Pandey, K.K., Sahoo, P.: A review on harvesting, oil extraction and biofuels production technologies from microalgae. Renew. Sustain. Energy Rev. **24**, 159–171 (2013)
33. Mayer, F.D., et al.: Revisiting energy efficiency, renewability, and sustainability indicators in biofuels life cycle: analysis and standardization proposal. J. Clean. Prod. **252**, 119850 (2020)
34. Omer, A.M.: Identifying, developing, and moving sustainable communities through application of bioenergy for energy or materials: future perspective through energy efficiency. Int. J. Life Sci. (IJLS). **1**(1), 9–39 (2017)
35. Osman, H., Shigidi, I., Elkhaleefa, A.: Optimization of sesame seeds oil extraction operating conditions using the response surface design methodology. Sci. Study Res. Chem. Chem. Eng. Biotechnol. Food Ind. **17**(4), 335 (2016)
36. Haji-Moradkhani, A., Rezaei, R., Moghimi, M.: Optimization of pulsed electric field-assisted oil extraction from cannabis seeds. J. Food Process Eng. **42**(4), e13028 (2019)
37. Ibrahim, S.M., et al.: Comparison of different methods for producing bio oil from Egyptian jatropha seeds. Biofuels. (2017)
38. Mondor, M.: Chia (Salvia Hispanica) seed oil extraction by-product and its edible applications. Food Rev. Intl. **40**(1), 115–134 (2024)
39. Kośmider, A., Leja, K., Czaczyk, K.: Improved utilization of crude glycerol by-product from biodiesel production. In: Biodiesel-Quality, Emissions and By-Products. Citeseer (2011)
40. Chew, C.L., et al.: Prospects of palm fruit extraction technology: palm oil recovery processes and quality enhancement. Food Rev. Intl. **38**(sup1), 893–920 (2022)
41. Marchetti, J.M.: A summary of the available technologies for biodiesel production based on a comparison of different feedstock's properties. Process Saf. Environ. Prot. **90**(3), 157–163 (2012)
42. Salvi, B., Panwar, N.: Biodiesel resources and production technologies–a review. Renew. Sustain. Energy Rev. **16**(6), 3680–3689 (2012)
43. Lin, L., et al.: Opportunities and challenges for biodiesel fuel. Appl. Energy. **88**(4), 1020–1031 (2011)
44. Gaurav, A., et al.: A kinetic model for a single step biodiesel production from a high free fatty acid (FFA) biodiesel feedstock over a solid heteropolyacid catalyst. Green Energy Environ. **4**(3), 328–341 (2019)
45. Helwani, Z., et al.: Solid heterogeneous catalysts for transesterification of triglycerides with methanol: a review. Appl. Catal. A. Gen. **363**(1–2), 1–10 (2009)

46. Zulqarnain, et al.: Comprehensive review on biodiesel production from palm oil mill effluent. ChemBioEng Rev. **8**(5), 439–462 (2021)
47. Venkataramana, S.H., et al.: Niger seed oil-based biodiesel production using transesterification process: experimental investigation and optimization for higher biodiesel yield using box–behnken design and artificial intelligence tools. Appl. Sci. **12**(12), 5987 (2022)
48. Bhatia, S.K., et al.: An overview on advancements in biobased transesterification methods for biodiesel production: oil resources, extraction, biocatalysts, and process intensification technologies. Fuel. **285**, 119117 (2021)
49. Kayode, B., Hart, A.: An overview of transesterification methods for producing biodiesel from waste vegetable oils. Biofuels. **10**(3), 419–437 (2019)
50. Ajith, B., et al.: Engine performance and exhaust emissions of Garcinia gummi-gutta based biodiesel–diesel and ethanol blends. SN Appl. Sci. **3**, 1–11 (2021)
51. Shen, W., et al.: A comprehensive review on the intricate processes involved in algae pyrolysis mechanism and possible migration of undesirable chemical elements. J. Anal. Appl. Pyrolysis, 106365 (2024)
52. Hu, B., et al.: Biofuels production using pyrolysis techniques. In: Advances in Biofuels Production, Optimization and Applications, pp. 103–125. Elsevier (2024)
53. Sharma, A.K., et al.: Holistic utilization of Chlorella pyrenoidosa microalgae for extraction of renewable fuels and value-added biochar through in situ transesterification and pyrolysis reaction process. Biomass Convers. Biorefinery. **14**(4), 5261–5274 (2024)
54. Kim, S.H., et al.: Hydroprocessing characteristics of palm fatty acid distillate in palm oil into low-carbon biofuel. Fuel. **364**, 131058 (2024)
55. de Paz Carmona, H., et al.: Effectiveness of Mo, NiMo, and CoMo catalysts for co-hydroprocessing furfural-acetone aldol condensation adducts with atmospheric gas oil to produce biofuels. Fuel. **355**, 129489 (2024)
56. Shalfoh, E., et al.: Fish waste oil extraction using supercritical CO2 extraction for biodiesel production: mathematical, and kinetic modeling. Renew. Energy. **220**, 119659 (2024)
57. Mubarak, M.: Supercritical CO2 extraction of lipid from microalgae for biodiesel production-a comprehensive review. SAJ Biotechnol. **9**, 101 (2024)
58. Arumugam, S., Baskar, S., Sriram, G.: Intensification of oil yield from Kokum seeds: an Energy-efficient Extraction using Ultrasonication Technique. Chem. Eng. Process Process Intensific. **196**, 109679 (2024)
59. Tabtimmuang, A., et al.: Ultrasonic-assisted synthesis of mono-and diacylglycerols and purification of crude glycerol derived from biodiesel production. Ind. Crop Prod. **208**, 117891 (2024)
60. Moschona, A., et al.: Optimization of enzymatic transesterification of acid oils for biodiesel production using a low-cost lipase: the effect of transesterification conditions and the synergy of lipases with different regioselectivity. Appl. Biochem. Biotechnol. **196**(11), 8168–8189 (2024)
61. Sharma, D., et al.: Enzymatic transesterification of waste cooking oil. In: Enzymes in Oil Processing, pp. 55–82. Elsevier (2024)
62. Chakraborty, D.D., Nath, L.K., Chakraborty, P.: Enzymatic biodiesel production. In: Enzymes in Oil Processing, pp. 15–31. Elsevier (2024)
63. Thippeshnaik, G., et al.: Experimental investigation of compression ignition engine combustion, performance, and emission characteristics of ternary blends with higher alcohols (1-heptanol and n-octanol). Energies. **16**(18), 6582 (2023)
64. Praveena, V., et al.: A systematic review on biofuel production and utilization from algae and waste feedstocks–a circular economy approach. Renew. Sustain. Energy Rev. **192**, 114178 (2024)
65. Awogbemi, O., Von Kallon, D.V.: Recent advances in the application of nanomaterials for improved biodiesel, biogas, biohydrogen, and bioethanol production. Fuel. **358**, 130261 (2024)

66. Li, K., Xie, W.: Enhanced biodiesel production from low-value acidic oils using ordered hierarchical macro–mesoporous MoAl@ H-SiO2 catalyst. Fuel. **364**, 131105 (2024)
67. Karimi, S., Saidi, M.: Application of nano hydrophobic sulfated mordenite as a novel catalyst for biodiesel production from neem seed-derived oil by electrochemical method. Energ. Conver. Manage. **299**, 117886 (2024)
68. Ruatpuia, J.V., et al.: Comparative life cycle cost analysis of bio-valorized magnetite nanocatalyst for biodiesel production: modeling, optimization, kinetics and thermodynamic study. Bioresour. Technol. **393**, 130160 (2024)
69. Sangeetha, B., Baskar, G.: Process design, kinetics, simulation, and techno-economic analysis of biodiesel production from Pongamia pinnata seed oil using a magnetically recyclable acidic ionic liquid catalyst. Energ. Conver. Manage. **301**, 118040 (2024)
70. Govindhan, P., et al.: Technologies of bio-diesel production: a state of the art review. Pet. Sci. Technol. **42**(3), 339–356 (2024)
71. Saidi, M., Amirnia, R.: Recent advances on application of metal-organic framework based catalysts in biodiesel production process: a review of catalyst types and activity, challenges and opportunities. Fuel. **363**, 130905 (2024)
72. Jiang, L., et al.: Mobilization of lipase and metal in microgel for biodiesel production and ring-opening polymerization. Surfaces Interfaces. **44**, 103783 (2024)
73. Thipdech, A., Prasertsit, K., Photaworn, S.: Process intensification of biodiesel production using pilot-scale continuous multiple baffle reactor with feed distribution. Chem. Eng. Process Process Intensific. **195**, 109614 (2024)
74. Gadore, V., Mishra, S.R., Ahmaruzzaman, M.: Advances in photocatalytic biodiesel production: preparation methods, modifications and mechanisms. Fuel. **362**, 130749 (2024)
75. Bansod, Y., et al.: Evaluating the environmental impact of crude glycerol purification derived from biodiesel production: a comparative life cycle assessment study. J. Clean. Prod. **437**, 140485 (2024)
76. Mohammadi, P., et al.: Quality control in biodiesel production plants. In: Sustainable Biodiesel, pp. 153–183. Elsevier (2024)
77. Hosseinzadeh-Bandbafha, H., et al.: Biodiesel plants: real-world sustainability analysis using environmental and social life cycle assessment. In: Sustainable Biodiesel, pp. 301–332. Elsevier (2024)
78. Selvam, C.D., Devarajan, Y., Jayabal, R.: Biodiesel production in India: prospects, challenges, and sustainable directions. Biotechnol. Bioeng. (2024)
79. Silva Jr., J.L., et al.: Smart scale-up of micromixers for efficient continuous biodiesel synthesis: a numerical study for process intensification. Chem. Eng. Process Process Intensific. **196**, 109664 (2024)
80. Miftah, F.K., Mutta, D.: Potential markets and policies for sustainable liquid biofuel production with emphasis to Eastern Africa countries: a review. Energy Sustain. Soc. **14**(1), 1 (2024)
81. Fukuda, H., Kondo, A., Noda, H.: Biodiesel fuel production by transesterification of oils. J. Biosci. Bioeng. **92**(5), 405–416 (2001)
82. Goff, M.J., et al.: Acid-catalyzed alcoholysis of soybean oil. J. Am. Oil Chem. Soc. **81**(4), 415–420 (2004)
83. Kim, H.-J., et al.: Transesterification of vegetable oil to biodiesel using heterogeneous base catalyst. Catal. Today. **93**, 315–320 (2004)
84. Meng, Y.-L., et al.: Effect of calcination temperature on the activity of solid Ca/Al composite oxide-based alkaline catalyst for biodiesel production. Bioresour. Technol. **128**, 305–309 (2013)
85. Pikula, K., et al.: The advances and limitations in biodiesel production: feedstocks, oil extraction methods, production, and environmental life cycle assessment. Green Chem. Lett. Rev. **13**(4), 275–294 (2020)
86. Singh, D., et al.: A review on feedstocks, production processes, and yield for different generations of biodiesel. Fuel. **262**, 116553 (2020)

87. Namasivayam, A., et al.: Biodiesel, emulsified biodiesel and dimethyl ether as pilot fuels for natural gas fuelled engines. Appl. Energy. **87**(3), 769–778 (2010)
88. Ziejewski, M., et al.: Diesel engine evaluation of a nonionic sunflower oil-aqueous ethanol microemulsion. J. Am. Oil Chem. Soc. **61**(10), 1620–1626 (1984)
89. Maher, K., Bressler, D.: Pyrolysis of triglyceride materials for the production of renewable fuels and chemicals. Bioresour. Technol. **98**(12), 2351–2368 (2007)
90. Xin, J., Imahara, H., Saka, S.: Oxidation stability of biodiesel fuel as prepared by supercritical methanol. Fuel. **87**(10–11), 1807–1813 (2008)
91. Mahmudul, H., et al.: Production, characterization and performance of biodiesel as an alternative fuel in diesel engines–a review. Renew. Sustain. Energy Rev. **72**, 497–509 (2017)
92. Palitsakun, S., et al.: Transesterification of Jatropha oil to biodiesel using SrO catalysts modified with CaO from waste eggshell. Catal. Commun. **149**, 106233 (2021)
93. Kaur, M., Malhotra, R., Ali, A.: Tungsten supported Ti/SiO2 nanoflowers as reusable heterogeneous catalyst for biodiesel production. Renew. Energy. **116**, 109–119 (2018)
94. Marchetti, J., Errazu, A.: Comparison of different heterogeneous catalysts and different alcohols for the esterification reaction of oleic acid. Fuel. **87**(15–16), 3477–3480 (2008)
95. Kalavathy, G., Baskar, G.: Synergism of clay with zinc oxide as nanocatalyst for production of biodiesel from marine Ulva lactuca. Bioresour. Technol. **281**, 234–238 (2019)
96. Roschat, W., et al.: Rubber seed oil as potential non-edible feedstock for biodiesel production using heterogeneous catalyst in Thailand. Renew. Energy. **101**, 937–944 (2017)
97. Baskar, G., Selvakumari, I.A.E., Aiswarya, R.: Biodiesel production from castor oil using heterogeneous Ni doped ZnO nanocatalyst. Bioresour. Technol. **250**, 793–798 (2018)
98. Bano, S., et al.: Fabrication and optimization of nanocatalyst for biodiesel production: an overview. Front. Energy Res. **8**, 579014 (2020)
99. Sudalai, S., et al.: Sustainable biodiesel production from Madhuca indica oil using a functionalized industrial waste as a catalyst: ready to scale-up approach. Ind. Crop Prod. **193**, 116233 (2023)
100. Ghani, N., et al.: Comparison of photo-esterification capability of bismuth vanadate with reduced graphene oxide bismuth vanadate (RGO/BiVO4) composite for biodiesel production from high free fatty acid containing non-edible oil. ChemistrySelect. **5**(29), 9245–9253 (2020)
101. Balat, M., Balat, H.: Progress in biodiesel processing. Appl. Energy. **87**(6), 1815–1835 (2010)
102. Mittal, V., Talapatra, K.N., Ghosh, U.K.: A comprehensive review on biodiesel production from microalgae through nanocatalytic transesterification process: lifecycle assessment and methodologies. Int. Nano Lett. **12**(4), 351–378 (2022)
103. Nandi, S., Bhattacharyya, R., Mandal, B.: Optimization and kinetics of biodiesel from jojoba (Simmondsia chinensis L. Schneider) oil. World Scientific News. **159**, 108–121 (2021)
104. Etim, A.O., Musonge, P., Eloka-Eboka, A.C.: Process optimization of bio-alkaline catalysed transesterification of flax seed oil methyl ester. Sci. Afr. **16**, e01275 (2022)
105. Negm, N.A., et al.: Biodiesel production from one-step heterogeneous catalyzed process of Castor oil and Jatropha oil using novel sulphonated phenyl silane montmorillonite catalyst. J. Mol. Liq. **234**, 157–163 (2017)
106. Ahmad, T., et al.: Optimization of process variables for biodiesel production by transesterification of flaxseed oil and produced biodiesel characterizations. Renew. Energy. **139**, 1272–1280 (2019)
107. Abdulrahman, A., Ali, A., Alfazazi, A.: Synthesis and process parameter optimization of biodiesel from Jojoba oil using response surface methodology. Arab. J. Sci. Eng. **46**(7), 6609–6617 (2021)
108. Asomaning, J., Mussone, P., Bressler, D.C.: Pyrolysis of polyunsaturated fatty acids. Fuel Process. Technol. **120**, 89–95 (2014)
109. Marx, S.: Glycerol-free biodiesel production through transesterification: a review. Fuel Process. Technol. **151**, 139–147 (2016)
110. Kartika, I.A., et al.: Biodiesel production from jatropha seeds: solvent extraction and in situ transesterification in a single step. Fuel. **106**, 111–117 (2013)

111. Baskar, G., et al.: Optimization and kinetics of biodiesel production from Mahua oil using manganese doped zinc oxide nanocatalyst. Renew. Energy. **103**, 641–646 (2017)
112. Farooq, M., et al.: Biodiesel production from date seed oil (Phoenix dactylifera L.) via egg shell derived heterogeneous catalyst. Chem. Eng. Res. Des. **132**, 644–651 (2018)
113. Noreen, S., et al.: Eco-benign approach to produce biodiesel from neem oil using heterogeneous nano-catalysts and process optimization. Environ. Technol. Innov. **22**, 101430 (2021)
114. Benguluri, S., et al.: Forward and reverse mappings in metal casting—a step towards quality casting and automation (11-009). AFS Trans. Am. Foundry Soc. **119**, 19 (2011)
115. Karmakar, B., et al.: Factorial optimization of biodiesel synthesis from castor-karanja oil blend with methanol-isopropanol mixture through acid/base doped Delonix regia heterogeneous catalysis. Fuel. **285**, 119197 (2021)
116. Chowdhury, S., et al.: Biodiesel synthesis from transesterified Madhuca indica oil by waste egg shell–derived heterogeneous catalyst: parametric optimization by Taguchi approach. Biomass Convers. Biorefinery. **11**, 1171–1181 (2021)
117. Balamurugan, S., et al.: Application of grey relational analysis in biodiesel production from linseed oil using novel eggshell catalyst. Mater. Today Proc. **45**, 1962–1969 (2021)
118. Esmaeili, H., Yeganeh, G., Esmaeilzadeh, F.: Optimization of biodiesel production from Moringa oleifera seeds oil in the presence of nano-MgO using Taguchi method. Int. Nano Lett. **9**(3), 257–263 (2019)
119. Mukherjee, I., Ray, P.K.: A review of optimization techniques in metal cutting processes. Comput. Ind. Eng. **50**(1–2), 15–34 (2006)
120. Jagdev, H., Browne, J., Keogh, J.: Manufacturing process optimisation—a survey of techniques. In: Proceedings of the Twenty-eighth International: Matador Conference. Springer (1990)
121. Niju, S., et al.: Optimization of acid catalyzed esterification and mixed metal oxide catalyzed transesterification for biodiesel production from Moringa oleifera oil. Green Process Synthes. **8**(1), 756–775 (2019)
122. Samuel, O.D., et al.: Grey wolf optimizer for enhancing Nicotiana Tabacum L. oil methyl ester and prediction model for calorific values. Case Stud. Therm. Eng. **35**, 102095 (2022)
123. Sai, B.A., et al.: Optimization of continuous biodiesel production from rubber seed oil (RSO) using calcined eggshells as heterogeneous catalyst. J. Environ. Chem. Eng. **8**(1), 103603 (2020)
124. Chhabra, M., Saini, B.S., Dwivedi, G.: Optimization of the dual stage procedure of biodiesel synthesis from Neem oil using RSM based Box Behnken design. Energy Sourc. Part A Recov. Util. Environ. Eff. **46**(1), 8888–8911 (2024)
125. Dwivedi, G., Sharma, M.: Application of Box–Behnken design in optimization of biodiesel yield from Pongamia oil and its stability analysis. Fuel. **145**, 256–262 (2015)
126. Kashyap, S.S., Gogate, P.R., Joshi, S.M.: Ultrasound assisted synthesis of biodiesel from karanja oil by interesterification: intensification studies and optimization using RSM. Ultrason. Sonochem. **50**, 36–45 (2019)
127. Rakić, T., et al.: Comparison of full factorial design, central composite design, and box-behnken design in chromatographic method development for the determination of fluconazole and its impurities. Anal. Lett. **47**(8), 1334–1347 (2014)
128. Kolakoti, A., et al.: Optimization and modelling of mahua oil biodiesel using RSM and genetic algorithm techniques. Math Models Eng. **6**(2), 134–146 (2020)
129. Jamshaid, M., et al.: Production optimization and tribological characteristics of cottonseed oil methyl ester. J. Clean. Prod. **209**, 62–73 (2019)
130. Tan, S.X., et al.: Two-step catalytic reactive extraction and transesterification process via ultrasonic irradiation for biodiesel production from solid Jatropha oil seeds. Chem. Eng. Process Process Intensific. **146**, 107687 (2019)
131. Lakshmi, S., et al.: Biodiesel production from rubber seed oil using calcined eggshells impregnated with Al. Braz. J. Chem. Eng. **37**, 351–368 (2020)

132. Bharadwaj, A.S., et al.: Biodiesel production from rubber seed oil using calcium oxide derived from eggshell as catalyst–optimization and modeling studies. Green Process Synthes. **8**(1), 430–442 (2019)
133. Al-Mawali, K.S., et al.: Life cycle assessment of biodiesel production utilising waste date seed oil and a novel magnetic catalyst: a circular bioeconomy approach. Renew. Energy. **170**, 832–846 (2021)
134. Samuel, O.D., Boye, T.E., Enweremadu, C.C.: Financial and parametric study of biodiesel production from hemp and tobacco seed oils in modified fruit blender and prediction models of their fuel properties with diesel fuel. Bioresour. Technol. Rep. **12**, 100599 (2020)
135. Arivarasu, A., et al.: Production of tobacco (Nicotiana tabacum L.) seed oil methyl esters using Tungstophosphoric acid as catalyst. In: International Conference on Advances and Innovations in Recycling Engineering. Springer (2021)
136. Khalil, I., et al.: Response surface methodology for the optimization of the production of rubber seed/palm oil biodiesel, IDI diesel engine performance, and emissions. Biomass Convers. Biorefinery. **7**, 37–49 (2017)
137. Ganapathy, T., Murugesan, K.A., Gakkhar, R.: Performance optimization of Jatropha biodiesel engine model using Taguchi approach. Appl. Energy. **86**(11), 2476–2486 (2009)
138. Khanjani, A., Sobati, M.A.: Performance and emission of a diesel engine using different water/waste fish oil (WFO) biodiesel/diesel emulsion fuels: optimization of fuel formulation via response surface methodology (RSM). Fuel. **288**, 119662 (2021)
139. Kumar, S., Jain, S., Kumar, H.: Experimental study on biodiesel production parameter optimization of jatropha–algae oil mixtures and performance and emission analysis of a diesel engine coupled with a generator fueled with diesel/biodiesel blends. ACS Omega. **5**(28), 17033–17041 (2020)
140. Elkelawy, M., et al.: Experimental studies on the biodiesel production parameters optimization of sunflower and soybean oil mixture and DI engine combustion, performance, and emission analysis fueled with diesel/biodiesel blends. Fuel. **255**, 115791 (2019)
141. Dubey, A., et al.: Optimization of diesel engine performance and emissions with biodiesel-diesel blends and EGR using response surface methodology (RSM). Clean. Eng. Technol. **8**, 100509 (2022)
142. Sivaramakrishnan, K., Ravikumar, P.: Optimization of operational parameters on performance and emissions of a diesel engine using biodiesel. Int. J. Environ. Sci. Technol. **11**, 949–958 (2014)
143. Parappagoudar, M.B., Vundavilli, P.R.: Application of modeling tools in manufacturing to improve quality and productivity with case study. Proc. Manufact. Syst. **7**(4), 193–198 (2012)
144. Dhingra, S., Bhushan, G., Dubey, K.K.: Multi-objective optimization of combustion, performance and emission parameters in a jatropha biodiesel engine using non-dominated sorting genetic algorithm-II. Front. Mech. Eng. **9**, 81–94 (2014)
145. Sekhar, S.C., et al.: Biodiesel production process optimization from Pithecellobium dulce seed oil: performance, combustion, and emission analysis on compression ignition engine fuelled with diesel/biodiesel blends. Energ. Conver. Manage. **161**, 141–154 (2018)
146. Singh, Y., et al.: Optimization of performance and emission parameters of direct injection diesel engine fuelled with pongamia methyl esters-response surface methodology approach. Ind. Crop Prod. **126**, 218–226 (2018)
147. Ansari, N.A., Sharma, A., Singh, Y.: Performance and emission analysis of a diesel engine implementing polanga biodiesel and optimization using Taguchi method. Process Saf. Environ. Prot. **120**, 146–154 (2018)
148. Dey, S., et al.: Study on performance-emission trade-off and multi-objective optimization of diesel-ethanol-palm biodiesel in a single cylinder CI engine: a Taguchi-fuzzy approach. Energy Sourc. Part A Recov. Util. Environ. Eff., 1–21 (2020)
149. Peter, A.S., et al.: Optimization of biodiesel production by transesterification of palm oil and evaluation of biodiesel quality. Mater. Today Proc. **42**, 1002–1007 (2021)

150. Zhang, Q., Ogren, R.M., Kong, S.-C.: A comparative study of biodiesel engine performance optimization using enhanced hybrid PSO–GA and basic GA. Appl. Energy. **165**, 676–684 (2016)
151. Jena, S.P., Mahapatra, S., Acharya, S.K.: Optimization of performance and emission characteristics of a diesel engine fueled with Karanja biodiesel using Grey-Taguchi method. Mater. Today Proc. **41**, 180–185 (2021)
152. Pohit, G., Misra, D.: Optimization of performance and emission characteristics of diesel engine with biodiesel using grey-taguchi method. J. Eng. **2013** (2013)
153. Shrivastava, K., Thipse, S., Patil, I.: Optimization of diesel engine performance and emission parameters of Karanja biodiesel-ethanol-diesel blends at optimized operating conditions. Fuel. **293**, 120451 (2021)
154. Etghani, M.M., et al.: A hybrid method of modified NSGA-II and TOPSIS to optimize performance and emissions of a diesel engine using biodiesel. Appl. Therm. Eng. **59**(1–2), 309–315 (2013)
155. Rao, K.P., Rao, B.A.: Parametric optimization for performance and emissions of an IDI engine with Mahua biodiesel. Egypt. J. Pet. **26**(3), 733–743 (2017)
156. Ayhan, V., et al.: Optimization of the factors affecting performance and emissions in a diesel engine using biodiesel and EGR with Taguchi method. Fuel. **261**, 116371 (2020)
157. Yessian, S., Varthanan, P.A.: Optimization of performance and emission characteristics of catalytic coated ic engine with biodiesel using grey-taguchi method. Sci. Rep. **10**(1), 2129 (2020)
158. Vellaiyan, S., Subbiah, A., Chockalingam, P.: Multi-response optimization to improve the performance and emissions level of a diesel engine fueled with ZnO incorporated water emulsified soybean biodiesel/diesel fuel blends. Fuel. **237**, 1013–1020 (2019)
159. Ganapathy, T., Gakkhar, R., Murugesan, K.: Optimization of performance parameters of diesel engine with Jatropha biodiesel using response surface methodology. Int. J. Sustain. Energy. **30**(sup1), S76–S90 (2011)
160. Singh, A., et al.: Optimization of performance and emission characteristics of CI engine fueled with Jatropha biodiesel produced using a heterogeneous catalyst (CaO). Fuel. **280**, 118611 (2020)
161. Shirneshan, A., Samani, B.H., Ghobadian, B.: Optimization of biodiesel percentage in fuel mixture and engine operating conditions for diesel engine performance and emission characteristics by Artificial Bees Colony Algorithm. Fuel. **184**, 518–526 (2016)
162. Gul, M., et al.: Grey-Taguchi and ANN based optimization of a better performing low-emission diesel engine fueled with biodiesel. Energy Sourc. Part A Recov. Util. Environ. Eff. **44**(1), 1019–1032 (2022)
163. Aydın, M., Uslu, S., Çelik, M.B.: Performance and emission prediction of a compression ignition engine fueled with biodiesel-diesel blends: a combined application of ANN and RSM based optimization. Fuel. **269**, 117472 (2020)
164. Veza, I., et al.: Grasshopper optimization algorithm for diesel engine fuelled with ethanol-biodiesel-diesel blends. Case Stud. Thermal Eng. **31**, 101817 (2022)
165. Shirneshan, A., et al.: Optimization and investigation the effects of using biodiesel-ethanol blends on the performance and emission characteristics of a diesel engine by genetic algorithm. Fuel. **289**, 119753 (2021)
166. Adam, I.K., et al.: Optimization of performance and emissions of a diesel engine fuelled with rubber seed-palm biodiesel blends using response surface method. Asian J. Appl. Sci. **4**(2) (2016)
167. Frey, D.D., Engelhardt, F., Greitzer, E.M.: A role for "one-factor-at-a-time" experimentation in parameter design. Res. Eng. Des. **14**, 65–74 (2003)
168. Yates, F., Mather, K.: Ronald Aylmer Fisher, 1890–1962. The Royal Society, London (1963)
169. Montgomery, D.C.: Design and Analysis of Experiments. John wiley & sons (2017)
170. Bhemuni, V., Chalamalasetti, S.: A review on hard turning by using design of experiments. J. Manufact. Sci. Prod. **13**(3), 209–219 (2013)

Chapter 3
Experimentation, Modelling, and Analysis of Biodiesel Conversion and Their Engine Performance and Emission Characteristics

3.1 Introduction

Renewable energy sources are potential alternatives to conventional diesel fuel that satisfy global energy demand and reduce environmental impact. The second-generation Garcinia Gummi-Gutta (GGG) feedstock grown on non-arable forest land ensures cost-effective feedstock useful for biodiesel production. This chapter discusses the details of the feedstock, oil extraction methods, and biodiesel production. The measurement procedures of fuel properties and their practicality in use in diesel engines were examined. The silver oxide nanoparticles were subjected to various characterizations, confirming their elemental composition and dimensions. The practicality of transforming crude oil to biodiesel conversion using suitable methods is highlighted based on examining oil properties. The procedure of the transesterification process towards biodiesel conversion was discussed. The transesterification experiments were designed based on influencing process variables on biodiesel yield, with a central composite design discussed. The detailed insights of individual and interaction effects of process variables were analysed, and empirical input-output relationships were established. The developed model prediction accuracy and optimized conditions (determined by metaheuristic algorithms) were tested experimentally and validated the model. The impact of silver oxide nanoparticles as a catalyst was examined on biodiesel conversion. The biodiesel production cost was estimated to correspond to optimized transesterification conditions. The biodiesel blends (diesel + biodiesel) were prepared and fuelled to diesel engines with different engine variables according to central composite design and analysed for their performance and emission characteristics. The impact of individual and interaction effects of engine variables with different blends was analysed with main effect, surface plots, and analysis of variance. The model's statistical adequacies and prediction capabilities with derived empirical equations were tested against random experimental cases. The detailed insights of the above information are discussed in subsequent sections.

M. Patel G. C. et al., *Biofuel Production, Performance, and Emission Optimization*, Springer Optimization and Its Applications 215,
https://doi.org/10.1007/978-3-031-84806-3_3

3.2 Details of Feedstock Material (Garcinia Gummi-Gutta: GGG) for Biodiesel Production

Garcinia Gummi-Gutta (GGG) tree is a tropical Guttiferae family species grown mainly in the forest (non-arable land) regions across the coastal belt (Kerala and Karnataka) and western ghats of southern India [1, 2]. GGG trees grow up to 25 m tall in the Nilgiris [1]. GGG seeds are of the non-food grain type, wherein the oil is extracted through various processing routes. The seed contains approximately 30–40% oil [1]. GGG is a miscellaneous stray spice tree wherein the parts such as fruits, barks, leaves, and roots possess phytochemical and biological characteristics (lipids, fatty acids, and nutraceuticals) best suited for distinguished applications resulting in high medicinal value [3]. Hydroxycitric acids ensure that the fruits (juice comprised of anti-scorbutic, anthelmintic, and cardiotonic properties) are potential resources to reduce weight loss [4]. Fruit rinds also possess commercial value in helping to prepare curries and garnish [5]. A higher proportion of Vitamin C in the fruit is best suited to cure heart diseases [6]. The resin content extracted from the tree can treat tumours, fevers, ulcers, and parasites [7]. Once the oil is taken out of the seeds, the cake leftover can be used to make bio-composites that can be used in a variety of applications in the industry [8]. GGG seed oil contains stearic and oleic acid, enabling researchers to use graphene oxide to prepare nanocomposite coatings (protecting material against a corrosive environment) best suited for industrial applications [9]. The fallen leaves ensure improved soil fertility resulting in good compost [10, 11]. In addition, glycerol that settles at the bottom during the oil extraction process has applications in pharmaceutical industries (for manufacturing soap and toothpaste), textile industries (for manufacturing leather), cosmetic industries, hair conditioner, and food industries [12, 13]. The study was conducted on the revenue generated by farmers from 19 different non-timber forest products [14].

41% of farmers collected fruits, followed by 14% each of barks and seeds, 9% each of leaves, arils, and roots, and 4% of farmers collected flowers due to their high commercial values and applications [14]. The maximum revenue generated for local people from collecting and selling 19 different non-timber forest products is Rs. 82,666 per annum for GGG, Rs. 18,540 per annum for Garcinia indica and Rs. 16,440/annum for Artocarpus lakoocha, and so on [14].

3.3 Details of the Process of GGG Seed Collection

Figure 3.1 shows the specifics of the GGG feedstock. As seen in Fig. 3.1a, b, GGG trees have rounded crowns and drooping branches and grow to a height of 5–25 m. The flowers in Fig. 3.1c appear in a cluster of 5–20 and are coloured white, dull-white, or greenish yellow.

Figure 3.1d shows the unripe fruit appearing greenish. After a few days, the unripe fruit turns into small pumpkin structured ripe fruit with a bright yellowish colour (refer to Fig. 3.1e). The ripped fruits possess 6–8 grooves with an average size of 5 cm

Fig. 3.1 Feedstock of Garcinia Gummi-Gutta: (**a**, **b**) tree, (**c**) flowers, (**d**) unripe fruit, (**e–g**) ripped fruit with a cross-sectional view, (**h**) aril, (**i**) dried seed separated from the fruit rind, and (**j**) seeds ready for biodiesel conversion

Fig. 3.2 Fruit rinds of GGG: (**a**) dried using smoke produced by wooden logs and (**b**) dried fruit rinds

in diameter. The ripe fruit weighs in the ranges between 100 and 120 g. The grown ripened fruits in forest lands are plucked by the local people and sliced, followed by drying (sun/air or hot air supplied through fire) to remove the presence of 10–15% of moisture content (refer to Fig. 3.2). The dried fruit rinds are presented in Fig. 3.2b.

A succulent aril surrounds the seeds, as shown in Fig. 3.1h. The dried fruit rind and the separated seeds are depicted in Fig. 3.1i, j. An average of 6–8 seeds is extracted from the ripe fruits. The seed measures approximately 3 cm in length and 1 cm in width.

3.4 Details of the Process of Oil Extraction from GGG Seeds

Mechanical pressing, solvent extraction, and boiling methods are widely applied for oil extraction from seed-based feedstocks. The selection of an appropriate method for oil extraction relies on the seed type and the desired quantity and quality of the oil.

3.4.1 *Boiling Method*

Seeds are boiled in water to release oil (initially reducing moisture followed by grinding seeds to form a paste and extracting, scooping, and drying the oil subjected to heat). The boiling method is typically used for sesame seeds possessing low oil content. The experimental set-up of the boiling method employed to extract the oil from the GGG feedstocks is presented in Fig. 3.3. The yield of oil is computed using the following expression presented in Eq. (3.1).

$$\text{Yield}_{\text{oil extraction}} = \frac{\text{Weight of oil}}{\text{Weight of seed}} \tag{3.1}$$

Fig. 3.3 Boiling of the GGG seed powder with water

3.4.2 Mechanical Pressing

The seeds are crushed by applying mechanical pressure (i.e. screw press) to extract oil from GGG seeds. Seeds are crushed with the two major components (rotating screw and cylindrical barrel) of the mechanical expeller (press). The mechanical screw press used for experimentation is presented in Fig. 3.4a. The length of the screw flight and the space between the screw shaft and barrel determine the seed compression. Oil is extracted from the seed by gradually increasing the pressure; the ejected oil and seed cake are collected at the barrel's perimeter (Fig. 3.4b). The solid waste and the oil are separated by being subjected to a filtration process. The oil yield in the mechanical press can be enhanced with increased mechanical pressure. The crude oil obtained from the mechanical screw press is presented in Fig. 3.4c.

1. The GGG seeds are air-dried in a sunny environment.
2. The dried seeds are ground to obtain powders.
3. The powder is filtered, packed, and placed in a round flask apparatus to carry out the process.
4. Hexane catalyst is added for the smooth flow of the solvent extraction process.
5. Hexane extracts the oil content from the seed powder subjected to heat treatment at 50–55 for 3–4 cycles.
6. The obtained oil and hexane mixture are treated at 70 °C to recover the maximum concentration of hexane.

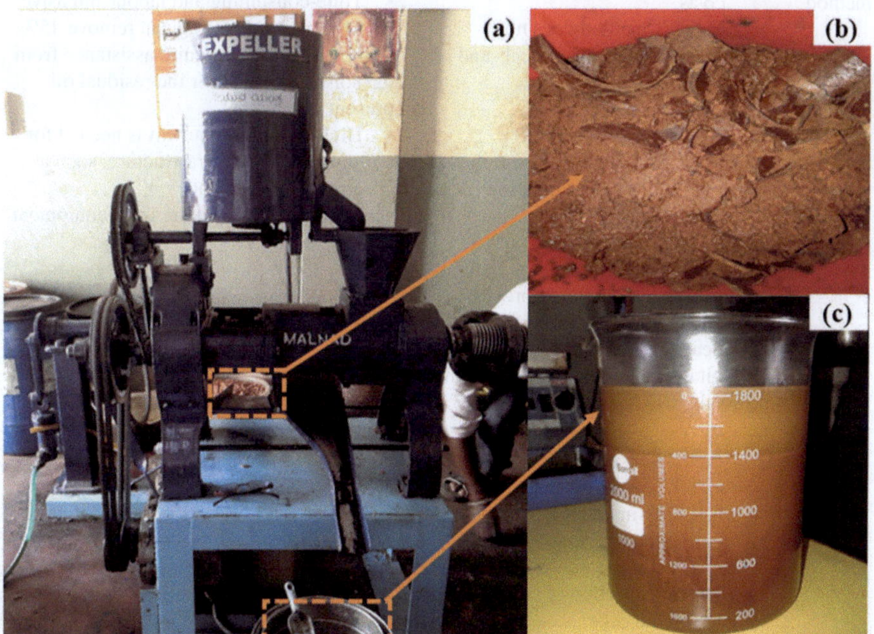

Fig. 3.4 Mechanical extraction of oil from GGG seeds: (**a**) Mechanical expeller, (**b**) cake, and (**c**) crude oil

Table 3.1 presents the details of oil extractions obtained from GGG feedstocks, including processing methods. The solvent extraction method yielded the highest oil recovery for GGG feedstock. The advantages and limitations discussed in Table 3.2 better explain the reasons.

Table 3.1 Comparison of oil yield (%) obtained from different extraction processes

	Oil yield (%)		
Feedstock	Boiling (%)	Mechanical expeller (%)	Solvent extraction (%)
GGG	18	23	42

Table 3.2 Advantages and limitations of oil extraction methods

Method	Advantages	Disadvantages
Boiling method	• Cost-effective method of producing oil • Requires minimal equipment and energy compared to other methods	• Tedious and time-consuming (energy-sapping, prone to drudgery, and inefficient) • Resulting in lower yield and quality • Boiling for a longer duration at higher temperatures results in the degradation of oil • Higher operating temperatures result in a higher risk, such as burns or injuries
Mechanical expression method	• Mechanical pressing results in lower operating or processing costs • This method does not require heat or chemicals and reduces the risk of contamination • The mechanical oil extraction method produces high-quality oil with a low concentration of FFAs • This method does not use heat, unlike the boiling method) for the oil extraction process, and therefore preserves heat-sensitive nutrients such as vitamins and antioxidants in the oil	• Large portion of oil is left in the cake or meal after extraction • Time-consuming and labour-intensive • Mechanical pressing can remove 45% of the oil and may require assistance from the solvent method for the residual oil present in the meal • High capital investment is needed for building essential infrastructure, such as specialized equipment • Risk of contamination from equipment
Solvent extraction method	• The solvent extraction method ensures a higher yield of oil conversion from seeds, resulting in an effective oil extraction process • This efficiency is due to the lower proportion of residual oil left in the cake	• Higher solvent cost result in costlier oil conversion than mechanical pressing • High capital investment and longer duration for oil extraction consume more energy • This is not an environmentally friendly procedure because volatile organic compounds are emitted into the atmosphere during the extraction process • The atmospheric release of volatile organic compounds

3.4.3 Solvent Extraction Method

The solvent extraction process is employed for seeds with low oil content and when maximum yield is desired. Hexane, ether (diethyl and petroleum), and ethanol are typical solvents used for oil extraction. Solvent extraction, such as the Soxhlet apparatus, offers advantages in recycling during the oil extraction process. The steps involved in oil extraction using solvent extraction methods are presented below (refer to Fig. 3.5).

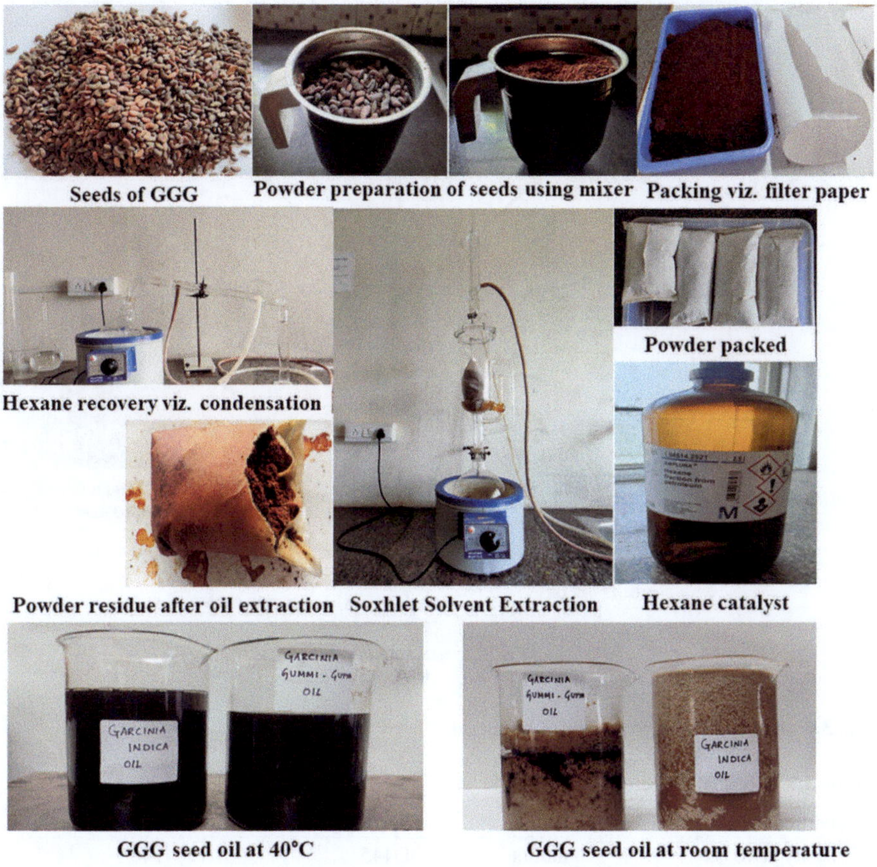

Fig. 3.5 Oil extraction process, viz., solvent extraction

3.5 Measurement Procedure for Various Characterization of Oils

After being extracted, seed oil has to be evaluated for fuel qualities, including density, viscosity, FFA, cloud point, cetane number, flash and fire points, molecular weight, and iodine value (see Fig. 3.6). ASTM criteria are followed in the inspection of these attributes. To determine the usefulness of diesel engines running on biodiesel, testing the fuel's characteristics is necessary. Table 3.3 lists the physical characteristics of GGG oil.

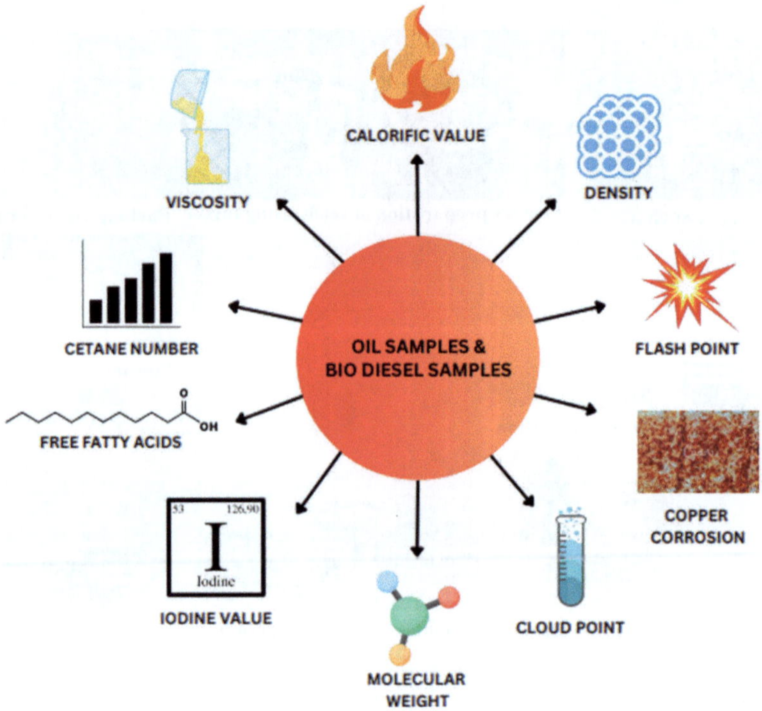

Fig. 3.6 Testing of fuel properties of biodiesel

Table 3.3 Physical properties of GGG oil

Properties	Units	Testing procedure ASTM	GGG oil
Viscosity at 40 °C	mm^2/s	D445	48
Density at 15 °C	kg/m^3	D93	882
Flash point	°C	D4052	246
Calorific value	MJ/kg	D240	38.800
Copper corrosion 50 °C, 3 h	–	D130	1a
Molecular weight	g/mol	D2502-87	886.23
Iodine value	g/100 g of oil	D5554	54.20
Free fatty acid	%	D664	13.56
Colour of oil at room temperature	–	–	Pale yellow

3.5.1 Viscosity

Viscosity is the measure of flow resistance in a fluid due to internal friction between molecules. A fluid with high viscosity is thick and offers greater flow resistance, while a fluid with low viscosity is thin and flows easily. For example, honey has a higher viscosity than water, making it thicker and slower to flow. The viscosity of seed oils and diesel can vary depending on the specific type being considered. Seed oils typically have higher viscosities due to their higher molecular weight and complex chemical structure compared to diesel fuel. The viscosity of a particular seed oil or diesel can be determined using a viscometer. It is important to note that the viscosity of a specific oil or diesel can change with temperature and pressure.

ASTM D445 standard test method is used to determine the kinematic viscosity of all samples. The test procedure includes recording the time for a set volume of fluid to flow through the capillary tube under gravity. The test conducted at a fixed temperature (e.g. 40 °C) and the viscosity is recoded in centistokes (cSt). The steps for conducting the viscosity measurements are as follows:

1. Take a measured quantity of the sample for testing.
2. Preheat the test sample to the desired temperature (e.g. 40 °C).
3. Fill the viscometer tube with the test sample.
4. Place the viscometer in a temperature-controlled bath and allow the sample to reach the desired temperature.
5. Record the time it takes for the sample to flow.
6. Repeat the test at least two times and record the average value.
7. Calculate the kinematic viscosity of the sample.

3.5.2 Density

The density of oils can be determined using hydrometers, pycnometers, and electronic density meters. The selection of a specific testing procedure relies on the type of oil and available equipment.

The present work employs hydrometer testing equipment, including a glass tube with a weighted bulb placed at the bottom with a scale on the stem. The hydrometer is immersed in the testing oil sample and determines the density by recording the scale where the oil surface intersects with the stem of the hydrometer. The oil or sample density is measured in g/mL or kg/m^3. ASTM D4052 is a standard test procedure used to determine the density of samples by a digital density meter. This test method applies to various liquids, including petroleum products, chemicals, and other organic liquids. The test involves filling a sample chamber with the liquid to be tested and measuring the mass and volume of the liquid using a digital density meter. The density is then calculated as the mass of the liquid divided by its volume.

The procedure for the ASTM D4052 test method is discussed below:

1. The measured quantity of the sample is taken for density measurement.
2. Degas the sample by vacuum or stirring to remove trapped air bubbles.
3. Fill the sample cell of the digital density meter with the test oils.
4. The oil sample to be tested is maintained at a set temperature between 15 °C and 30 °C.
5. Measure the density of the sample using the digital density meter, which involves the following steps:
   - Verify the instrument calibration and make any necessary adjustments.
   - Zero the instrument with air or water, depending on the instrument design.
   - Insert the sample cell into the instrument and record the density reading.
   - Repeat the measurement at least two more times to obtain an average value.
6. Calculate the relative density (specific gravity) of the sample using the formula presented in Eq. (3.2):

$$\text{Relative density} = \frac{\text{Density of sample}}{\text{Density of water}} \tag{3.2}$$

3.5.3 *Flash Point*

The flash point is the lowest temperature at which the vapours of test samples will ignite in the presence of an ignition source, such as a flame or spark. ASTM D93 standards were used to test the flash point where the samples were heated in a closed container and an ignition source was periodically introduced to vapour above the liquid. The temperature at which the flash is observed is recorded as the flash point temperature of the sample. The Tag Closed Cup (TCC) method is used to test samples with a flash point below 175 °C, while the Pensky-Martens Closed Cup (PMCC) method is used for samples with the flash point above 40 °C. The PMCC method is used for testing the flash point of the samples.

Samples with low flash points are more volatile and have a higher fire and explosion risk, while samples with high flash points are less volatile and have a lower fire and explosion risk. The flash point is an important property used in the classification and regulation of flammable and combustible liquids, as well as in developing safety guidelines and procedures for their handling and storage. The steps for performing the test are as follows:

- Take a metered quantity of the sample for flash point temperature measurement.
- Ensure that the testing samples are free of impurities (filter the samples if necessary).
- Fill the closed cup apparatus with the sample up to the specified level between 50 and 75% of the total volume.
- Insert the thermometer and the ignition source into the closed cup apparatus.

- Heat the sample at a specified rate between 1 and 5 °C per min.
- Observe the sample for the appearance of a flash of fire upon ignition.
- Record the temperature at which the flash occurs.

3.5.4 Cloud Point

Cloud point is a physical property of petroleum products or other liquids where the temperature at which solid or waxy particles begin to form causes the liquid to become cloudy or hazy. The measurement of the cloud point of fuels defines their practicality for use in diesel engines under cold environmental conditions. The cloud point affects the flow properties and performance in diesel engines. Fuels with high cloud points can cause fuel filter clogging and other problems. ASTM D2500 is a standard test method used to determine the cloud point of petroleum products, such as diesel fuels and heating oils. The procedure for the ASTM D2500 test method involves cooling a sample of the petroleum product at a specified rate and recording the temperature at which the first sign of cloudiness or haze appears. The steps for performing the test are as follows:

1. The metered quantity of sample is taken for cloud point temperature measurement.
2. Place the sample (free from impurities) in a cylindrical container with a tightly fitting lid.
3. Cool the sample at a controlled rate of 1–2 °C per min using a cooling bath or refrigeration equipment.
4. Observe the sample periodically for the first sign of cloudiness or haze, which is usually indicated by the appearance of a faint ring or spot.
5. Record the temperature at which the cloudiness or haze first appears.

3.5.5 Calorific Value

Calorific value is the measure of the amount of heat released when a fuel is burnt under a controlled environment, namely a calorimeter. It is a fuel's heating value or energy content expressed in units of energy per unit mass or volume. The two commonly used measures of calorific value are higher heating values (measured as heat released due to condensation of water vapour produced during the combustion process) and lower heating values (which do not include this heat and assumes the water vapour is fully evaporated and removed from the combustion products). Calorific value is an important property of fuels as it provides information on the fuel's energy content and potential heat output. It determines the efficiency of fuel combustion processes and helps to compare the energy content of different fuels. The calorific value of a fuel is affected by its chemical composition and factors such as moisture content and combustion conditions. ASTM D240 is a standard test method used to determine the heat of combustion of samples. The test involves burning a measured quantity of fuel in a bomb calorimeter and measuring the temperature change of the surrounding water. Calorific value is computed based on the mass of the fuel burnt and the temperature change of the water.

The procedure for estimating the calorific value of a fuel sample using a bomb calorimeter involves the following steps:

1. Prepare the sample: The sample is typically weighed and placed into a small, sealed container called a crucible. Ensure the sample is finely ground and homogenized so that it will burn completely and uniformly.
2. Assemble the bomb calorimeter: A bomb calorimeter typically consists of a steel bomb (a sealed container capable of withstanding high pressure and temperature), a water jacket surrounding the bomb, an ignition wire or fuse, a pressure gauge, and a thermometer. The bomb is filled with a measured amount of oxygen, and the water jacket is filled with a known amount of water.
3. Measure the initial temperature: Before the sample is burnt, the temperature of the water in the calorimeter is measured and recorded.
4. Ignite the sample: The crucible containing the sample is inserted into the bomb, and the bomb is sealed. An ignition wire or fuse ignites the sample, which burns entirely in the oxygen-filled bomb.
5. Measure the final temperature: The water in the calorimeter is measured again after the sample has burnt completely. The initial and final temperature difference is used to calculate the amount of heat released during the combustion reaction.
6. Calculate the calorific value: The calorific value of the sample can be calculated using the following formula:

$$\text{Calorific value} = \frac{\text{Heat released}}{\text{Mass of sample in crucible}} \tag{3.3}$$

7. Heat released is calculated as the product of the mass of water, the specific heat of water, and the temperature difference.

3.5.6 *Copper Corrosion*

The copper corrosion test measures the corrosive properties of petroleum products, including oil samples and diesel. The test evaluates the product's tendency to corrode copper and copper alloys when in contact with the product. The copper corrosion test evaluates the petroleum products exposed to corrode equipment and pipelines during transportation, storage, and use. Corrosion can lead to damage, leaks, and contamination of the product, as well as increased maintenance and repair costs.

ASTM D130 is a standard test method used to evaluate the corrosiveness of gasoline towards metals, specifically copper. The test method involves immersing a polished copper strip in a gasoline sample at a specified temperature (e.g. 50 °C) for a specific period (e.g. 3 h). After the immersion period, the copper strip is examined for signs of corrosion, such as discoloration, tarnishing, or pitting. The degree of corrosion is graded on a scale from 1 to 4, with one indicating no corrosion and four indicating severe corrosion.

The procedure involves the following steps:

1. Obtain a clean copper strip: The copper strip used in the test should be approximately 10 cm long and 1 cm wide, with a thickness of about 1.5 mm. The strip should be cleaned using a suitable solvent, such as acetone or ethanol, and dried before use.
2. Apply the sample to the copper strip: A small amount of the sample being tested (typically 5–10 mL) is placed on the copper strip, covering an area of approximately 10 cm^2.
3. Heat the copper strip: With the sample applied, the copper strip is heated in an oven at 50 °C for 3 h.
4. Inspect the copper strip: Upon heating, the copper strip is taken out from the oven and allowed to cool at room temperature. The strip is then visually inspected for signs of corrosion, such as discoloration, pitting, or etching.
5. Compare the results to a reference sample: To determine whether the sample being tested is corrosive, the results are compared with a known level of corrosiveness. The reference sample is typically a mixture of naphthenic acid and a solvent. It is assigned a rating of 1A (non-corrosive) to 4B (severely corrosive) based on its level of corrosion.
6. Report the results: The test results are the rating of the sample being tested based on the severity of corrosion observed. A rating of 1A indicates that no corrosion was observed while a rating of 4B indicates severe corrosion.

3.5.7 Molecular Weight

The molecular weight of oil samples varies depending on their chemical composition, as it affects physical and chemical behaviour such as viscosity, boiling point, and reactivity. Molecular weight determination is complex due to the wide range (100–1000 g per mole) of hydrocarbons and other compounds. Mass spectrometry, gas chromatography, and infrared spectroscopy are widely applied methods to determine oil samples' chemical composition and molecular weight.

ASTM D2502-87 is a standard test method to determine how the average molecular weight of petroleum products using vapour pressure osmometry. This method involves dissolving a sample of the petroleum product in a solvent, such as cyclohexane or toluene, and measuring the osmotic pressure (the pressure required to limit the flow of solvent across the semipermeable membrane that separate the solution from pure solvent) of the resulting solution. By measuring the osmotic pressure and estimating the molecular weight of the solvent, the average molecular weight of the petroleum product can be calculated using the Van't Hoff equation.

The procedure to determine the molecular weight of oil samples involves the following steps:

1. Prepare the sample: The sample should be a representative fraction of the tested oil, with a boiling point range appropriate for the chromatographic column being used. The sample should be dissolved in a suitable solvent, such as n-pentane or n-hexane, at approximately 10 mg/mL.
2. Inject the sample into the chromatograph: The sample is injected into the gas chromatograph using a suitable injection technique such as split or on-column injection.
3. Run the chromatograph: The chromatograph is run under conditions appropriate for the tested sample, using a suitable chromatographic column and detector. The column should have a non-polar stationary phase, such as silicone or polydimethylsiloxane.
4. Determine the retention time with the known molecular weight.
5. Calculate the molecular weight: Computation of the molecular weight of the sample is done using the Mark-Houwink equation or a similar relationship between molecular weight and retention time.

3.5.8 *Iodine Value*

The iodine value (ID) is a measure of oil unsaturation or the degree to which it is reactive with iodine. It refers to the number of iodine grams absorbed in 100 g of oil. Higher levels of unsaturated fatty acids result in higher iodine values, while higher levels of saturated fatty acids cause lower iodine values. The measurement of ID is important as it provides detailed insight into stability and susceptibility to oxidation. Oils with higher ID, such as soybean and linseed oil are often used in the production of paints, varnishes, and other industrial applications. On the other hand, oils with lower iodine values, like coconut and palm oils, are more stable and resistant to rancidity, making them suitable for use in food products.

The ASTM D5554 standards are used to determine the iodine value of fats and oils through the following steps:

1. The measured quantity of oil sample weighing between 0.2 and 0.5 g is used to record the oil sample's iodine value.
2. Dissolve the oil in a suitable solvent. Common solvents include chloroform, carbon tetrachloride, and a mixture of glacial acetic acid and chloroform. The solvent should be chosen to ensure the oil dissolves completely and does not react with the iodine reagent.
3. The measured quantity of oil sample weighing between 0.2 and 0.5 g is used to record the oil sample's iodine value.

4. Dissolve the oil in a suitable solvent. Common solvents include chloroform, carbon tetrachloride, and a mixture of glacial acetic acid and chloroform. The solvent should be chosen to ensure the oil dissolves completely and does not react with the iodine reagent.
5. Add iodine reagent: The iodine reagent is typically a solution of iodine and iodide in a chloroform solvent. The reagent is added in excess to ensure that all of the unsaturated bonds in the oil are reacted.
6. Titrate the excess iodine with a standard solution of sodium thiosulfate, referred to as concentration. The endpoint of the titration is typically detected using a starch indicator, which forms a blue-black complex with excess iodine.
7. Calculate the iodine value based on the number of grams of iodine absorbed by 100 g of oil. This is typically determined using Eq. (3.4).

$$\text{Iodine value} = \frac{\left[\begin{pmatrix}\text{Vol.of sodium thiosulfate solution used to titrate the sample}\\ \text{Vol.of sodium thiosulfate solution used to titrate the blank}\end{pmatrix}\times \\ \text{Normality of sodium thiosulfate solution} \times 127\right]}{\text{Weight of oil sample}} \quad (3.4)$$

In Eq. (3.4), 127 denotes the molecular weight of iodine.

3.5.9 Determination of FFA

The presence of FFA in seed oil is an essential factor for assessing the quality of seed oils. FFA is determined to know the oil's acid content, which signifies the degree of hydrolysis or rancidity. The FFA content of seed oils can be determined using the following methods.

1. **Acid-base titration:** This method involves titrating a sample of the oil with a standardized alkaline solution (such as NaOH) to neutralize the FFA. FFA content is computed based on the amount of alkaline solution used.
2. **The colorimetric** method involves reacting an oil sample with a colorimetric reagent (such as phenolphthalein) to produce a coloured complex. The intensity of the colour is then measured using a spectrophotometer, and the FFA content is calculated based on a standard curve.
3. **Enzymatic method:** This method uses enzymes (such as lipase) to hydrolyse the FFA in the oil selectively. The released fatty acids are then measured using a colorimetric or titration method.
4. **Infrared (IR) spectroscopy:** This method involves measuring the absorbance of the oil at a specific wavelength using an IR spectrophotometer. The absorbance at a particular wavelength is proportional to the FFA content, calculated based on a standard curve.

An acid-base titration is used because of its simplicity and low cost. FFA content of GGG seed oil is calculated using the following procedure:

Step 1: Preparation of N/10 NaOH Standard Solution

- Open the NaOH container and weigh about 4 g of NaOH using the weighing scale.
- The NaOH container should be stored in a dry place after usage.
- 1 L of distilled water is measured and taken in a 1 L standard conical flask.
- Then weigh 4 g of NaOH is added to the distilled water flask.
- The NaOH flakes are dissolved completely by continuous stirring using a glass rod.
- Make sure that the NaOH flakes are entirely dissolved in distilled water.
- Now the N/10 NaOH solution is ready.

Step 2: Calculating FFA Level of Oils by Titration

- The burette is filled with 25 mL of N/10 NaOH solution in the first step.
- In a 250 mL conical flask, 50 mL of Isopropyl alcohol is added.
- The solution in the flask is mixed with a few drops of N/10 NaOH solutions and shaken well.
- For the above explanation, 10 g of a metered quantity of prepared oil is added and shaken well.
- The above solution is heated to a temperature maintained at 60 °C.
- Shake the heated solution to a uniform mixture and cool to room temperature.
- Add 3–4 drops of Phenolphthalein Indicator and titrate with N/10 NaOH solution. Shake it vigorously until a pale pink colour (Fig. 3.7) appears, which should persist for at least 1 min. Record burette reading.

Fig. 3.7 The mixture in the conical flask turns a pale pink colour

Step 3: FFA Calculation

FFA computation is done using Eq. (3.5). The value 28.2 is computed based on the ratio of the molecular weight of oleic acid (282) and the value 10.

$$\text{FFA} = \frac{\left[28.2 \times \text{normality of NaOH} \times \text{titrate value}\right]}{\text{Weight of oil}} \tag{3.5}$$

In Eq. (3.5), the titrant value is the concentration of NaOH solution (in mL) consumed during titration.

The FFA level may vary with different oil extraction methods. FFA measures the quality of oil and reflects the level of free fatty acids present in the oil. The FFA level can increase with specific oil extraction methods that involve heat or solvents. For example, oils extracted using solvent-based methods such as hexane extraction or high-temperature processes may have a higher FFA level than those extracted using mechanical pressing methods (cold or expeller). Mechanical pressing methods typically involve less heat and pressure, which can help preserve the oil's quality and keep the FFA level low. However, these methods may also reduce oil yields and production costs. Oil extraction methods directly impact the FFA level, and their proportion varies with different methods. Therefore, the present work uses the solvent extraction method for oil extraction from GGG seeds. The unit of measurement for FFA is usually expressed as a percentage of the oil's weight or as a percentage of the oil's acidity. For example, a 2% FFA level means that 2% of the weight of the oil is made up of FFA. Another way to express this is to say that the oil has an acidity of 2%. The FFA content in GGG oil computed using Eq. (3.6).

$$\text{FFA}_{\text{GGG oil}} = \frac{\left(28.2 \times 0.1 \times 48\right)}{10} = 13.56\% \tag{3.6}$$

3.5.10 Cetane Number

The cetane number (CN) is a measure of the ignition quality (of diesel fuel, indicating how easily the fuel ignites and burns smoothly). It measures the speed at which the fuel ignites when sprayed into the compressed air in a diesel engine. The CN is calculated by comparing the ignition delay period of the fuel to that of a reference fuel. This comparison is done using reference fuels such as n-hexadecane and 1-methylnaphthalene, which have known cetane numbers. The CN of diesel fuel is determined using the formula provided in Eq. (3.7).

A higher cetane number indicates that the fuel ignites more efficiently and burns more smoothly, resulting in less engine noise and smoke and better overall performance. Diesel engines typically operate best with a cetane number between 40 and 55. Fuels with a cetane number below 40 may cause engine knock, increased emissions, and reduced fuel efficiency. Fuels with a cetane number above 55 may not provide any additional benefits and may be more expensive to produce. Fuel producers use CN to ensure fuel quality and performance. In many countries, it is also a regulatory requirement to meet minimum cetane number specifications for diesel fuel.

CN of diesel fuels and oils is determined by applying the ASTM D613 standard. The test involves injecting a fuel sample into a test engine and recording the time interval between the start of injection and the beginning of combustion. CN is then calculated based on this time interval, compared to the time intervals of standard reference fuels with known cetane numbers.

The procedure involves the following steps:

1. Sample preparation: A representative diesel fuel sample is collected and stored in a clean, dry container.
2. Test apparatus preparation: The test apparatus, which typically consists of a test engine, fuel injection system, and other related components, is prepared according to the manufacturer's instructions and applicable standards.
3. Test procedure: The diesel fuel sample is injected into the test engine, and the ignition delay time is measured. This is typically done using specialized equipment that can accurately measure the time between fuel injection and the start of combustion.
4. Calculation of cetane number: CN is calculated based on the ignition delay time using a standard equation or lookup table provided by the manufacturer or a relevant standard. The cetane number is typically reported as a numerical value, with higher values indicating better ignition quality.

$$CN = \left[\left(1\ \ MF\right) \times 100\right] + \left(MF \times CNMNT\right) \quad (3.7)$$

MF is the volume fraction of the cetane reference fuel required to match the ignition delay period of the sample, and CNMNT is the cetane number of the 1-methylnaphthalene reference fuel.

3.5.11 Characterization of Oil Sample Using GC-MS

The GC-MS analysis of GGG oil is performed at the Sophisticated Analytical Instruments Facility (SAIF)—IIT Madras. The analysis is conducted with multidimensional gas chromatography coupled with a GC-MS (Made-Shimazu, Japan) and non-polar and polar double capillary columns (30 m × 250 μm × 0.25 μm). High-purity helium was used as the carrier gas at a constant 1.2 mL/min flow rate. 1 μL of the sample (split ratio 15:1) is injected into the GC and GC-MS. The initial temperature is set to 75 °C with a pressure of 11.367 psi, heated to 280 °C, and maintained isothermally. The ion source temperature was set to 150 °C, the interface temperature to 230 °C, and the solvent delay time to 3 min. Electron impact (EI) ionization mode is set to 70 eV, and the column's linear velocity was 40.402 cm/s. The individual components are identified by comparing their mass spectra to those in the NIST library mass spectra database and those in the scientific literature.

The presence of numerous chemicals as mapped from the NIST library was shown by GC-MS analysis of GGG oil. The GGG oil gas chromatogram revealed several peaks with varying retention periods (Fig. 3.8). The results showed the presence of many compounds with the prominent ones including N-α, N-ω-Di-cbz-L-arginine; Tetradecane, 2, 6, 10-trimethyl-; Benzene, (1-methyl-dodecyl)-; cis-Vaccenic acid; 1-cis-Vaccenoylglycerol; Stearic anhydride; 9-Octadecenoic acid (Z)-, oxiranylmethyl ester; Trilinolein; Ethyl iso-allocholate; Oleic acid, eicosyl ester. The presence of cis-Vaccenic acid, 1-cis-Vaccenoylglycerol, and 9-Octadecenoic acid (Z)-, oxiranylmethyl ester in GGG oil is more compared to other compounds, which contribute to 54.15% of the total peak area of the chromatogram (Table 3.4).

Cis-Vaccenic acid, or 11-octadecenoic acid, is a monounsaturated omega-7 fatty acid. Cis-Vaccenic acid is considered to have potential health benefits, including anti-inflammatory properties and the ability to help regulate cholesterol levels. 1-cis-Vaccenoylglycerol is an ester formed by the combination of cis-Vaccenic acid with glycerol. 9-Octadecenoic acid (Z)-, oxiranylmethyl ester is a fatty acid ester that consists of 9-octadecenoic acid (also known as oleic acid) esterified with an oxiranylmethyl (epoxy methyl) group. Oleic acid is a monounsaturated omega-9 fatty acid.

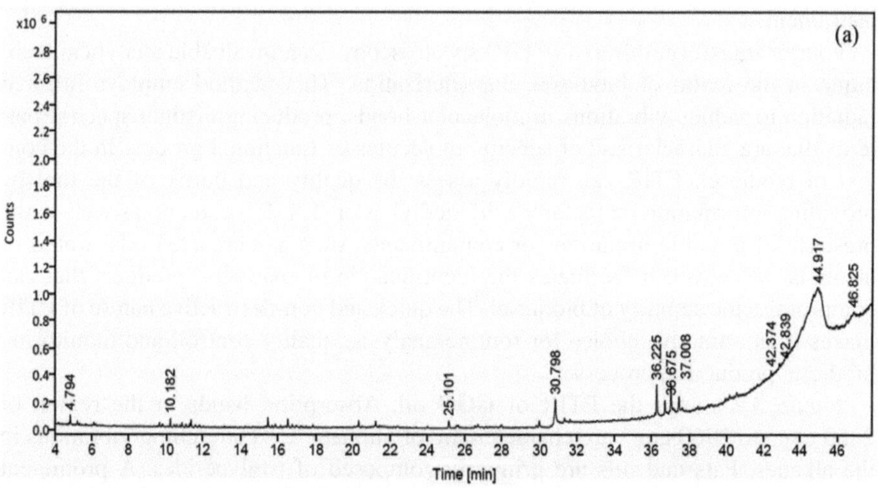

Fig. 3.8 GC-MS characterization of GGG oil

Table 3.4 Compounds identified in GGG oil by GC-MS

Peak	Retention time (min)	Compound name	Peak area (%)	Molecular formula	Molecular weight (g/mol)
1	4.794	N-α, N-ω-Di-cbz-L-arginine	5.09	$C_{22}H_{26}N_4O_6$	442.50
2	10.182	Tetradecane, 2, 6, 10-trimethyl	3.67	$C_{17}H_{36}$	240.50
3	25.101	Benzene, (1-methyldodecyl)-	4.98	$C_{17}H_{28}$	232.40
4	30.798	cis-Vaccenic acid	22.86	$C_{18}H_{34}O_2$	282.50
5	36.225	1-cis-Vaccenoylglycerol	18.26	$C_{21}H_{40}O_4$	356.00
6	36.675	Stearic anhydride	8.51	$C_{36}H_{70}O_3$	550.90
7	37.008	9-Octadecenoic acid (Z)-, oxiranylmethyl ester	13.03	$C_{21}H_{38}O_3$	338.50
8	42.374	Trilinolein	2.27	$C_{57}H_{98}O_6$	879.40
9	42.839	Ethyl iso-allocholate	2.20	$C_{26}H_{44}O_5$	436.63
10	44.917	Trilinolein	9.79	$C_{57}H_{98}O_6$	879.40
11	46.825	Oleic acid, eicosyl ester	9.34	$C_{38}H_{74}O_2$	562.99

3.5.12 Characterization of Oil Sample Using FTIR

The GGG oil was were characterized using the FTIR method. This test was conducted at the SAIF, established at IIT Bombay, with the support of the DST, Government of India, New Delhi. Table 3.5 shows the specifications of the FTIR instrument.

Fourier-transform infrared (FTIR) spectroscopy is an invaluable analytical technique in the realm of biodiesel characterization. This method employs infrared radiation to induce vibrations in molecular bonds, producing distinct spectral patterns that are characteristic of specific molecules or functional groups. In the context of biodiesel, FTIR can rapidly assess the quality and purity of the fuel by providing information on its fatty acid methyl ester (FAME) content, as well as the presence of possible impurities or contaminants, such as unreacted oils, water, or alcohols. Moreover, it facilitates the identification of oxidation products that can compromise the stability of biodiesel. The quick and non-destructive nature of FTIR makes it an attractive choice for routine analysis, quality control, and monitoring biodiesel production processes.

Figure 3.9 shows the FTIR of GGG oil. Absorption bands in the region of 2850 cm^{-1} to 2960 cm^{-1} are characteristic of aliphatic C-H stretching vibrations in the alkanes. Fats and oils are primarily composed of triglycerides. A prominent absorption band in the region of 1700 cm^{-1} to 1750 cm^{-1} is indicative of the carbonyl stretch (C=O) in esters. Absorption bands around 1000 cm^{-1} to 1300 cm^{-1} can be indicative of C-O stretching vibrations, typical of esters in triglycerides. The

Table 3.5 Specifications of FTIR instrument

Make	Bruker, Germany
Model	3000 hyperion microscope with vertex 80 FTIR system
Specification	• Focal plane array: 128 × 128, range: 4000–900 cm^{-1} • Single point detector: Range: 4000–450 cm^{-1} • Analysis area: 128 × 28 in 2D format on the sample plane 300 × 300 μm • Accessories: Micro ATR, grazing angle • Spatial resolution with 15× objective–2.7 μm • 20x objective (ATR): 0.5 μm • Temperature-controlled sample stage • Spectral resolution of FTIR 0.2 cm^{-1} • Rapid scan and step scan available. Rapid scan 65 spectra/s at 16 cm^{-1} • Library search for organic compounds and polymers

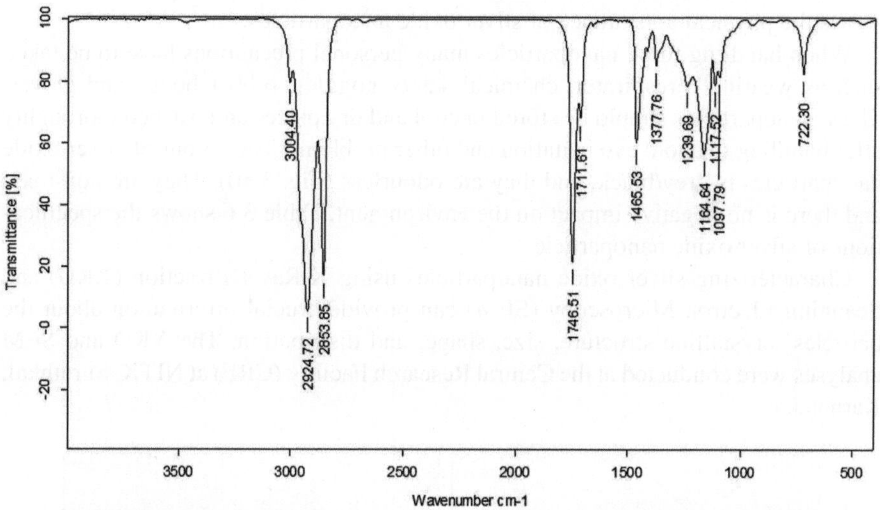

Fig. 3.9 FTIR of GGG oil

region below 1500 cm^{-1} is often referred to as the fingerprint region. This region can be complex but can offer additional insight into the specific types of molecules present.

3.6 Discussions on XRD and SEM Morphology of Ag_2O Nanoparticles

The nano additives are particles that are typically less than 60 nm in size. There are different types of nanoparticles available such as Silver Oxide, Manganese Oxide, Copper Oxide, and Cerium Oxide. Silver oxide (Ag_2O) nanoparticles possess a wide range of industrial applications including catalysis for chemical reactions [15], medical and healthcare applications due to their antibacterial properties [16], sensors due to their excellent sensitivity to electrical conductivity [17], photocatalytic activity for degrading organic pollutants and dyes in wastewater treatment [18], optoelectronics for enhanced light absorption and emission characteristics [19], biomedical applications [20], water purification [21], and food packaging [22]. The silver oxide nanoparticles have a higher surface area-to-volume ratio and electronic properties, ensuring the highest catalytic activity and enabling a faster and increased transesterification reaction rate that converts triglycerides to biodiesel [23]. Additionally, Ag_2O nanoparticles exhibit better thermal and chemical stability [24, 25] and reusability characteristics [26]. These characteristics make Ag_2O nanoparticles are useful for catalytic activity, providing an effective catalyst for biodiesel conversion. Silver oxide is used as a nano additive in the blends of GGG oil biodiesel. It is purchased from Intelligent Materials Pvt. Ltd., Punjab. Figure 3.10 shows the physical appearance of silver oxide nanoparticles.

When handling these nanoparticles many personal precautions have to be taken such as wearing a respirator, chemical safety goggles, rubber boots, and gloves. These nanoparticles should be stored in cool and dry places and washed thoroughly after handling to avoid eye irritation and other problems. The colour of silver oxide nanoparticles is grey/black, and they are odourless (Fig. 3.10). They are non-toxic and there is no negative impact on the environment. Table 3.6 shows the specifications of silver oxide nanoparticle.

Characterizing silver oxide nanoparticles using X-Ray Diffraction (XRD) and Scanning Electron Microscopy (SEM) can provide crucial information about the particles' crystalline structure, size, shape, and distribution. The XRD and SEM analyses were conducted at the Central Research Facility (CRF) at NITK, Surathkal, Karnataka.

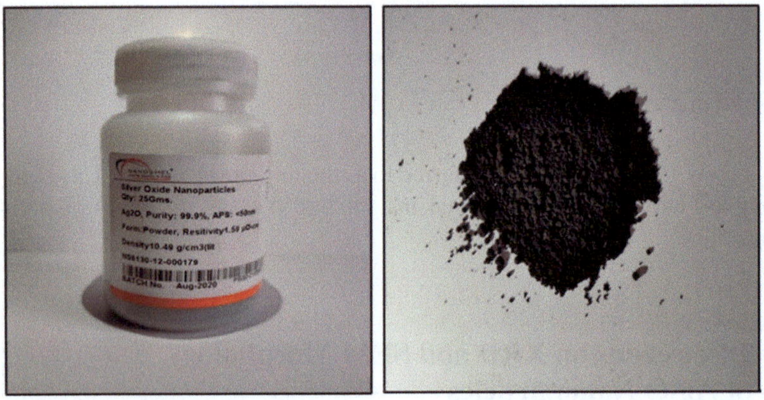

Fig. 3.10 Silver oxide nanoparticle

Table 3.6 Specification of silver oxide nanoparticle

Product	Silver oxide nanoparticle
Purity	99.9%
Molecular formula	Ag_2O
Molecular weight	231.74 g/mol
Colour	Grey/black
Form	Powder
Density	10.49 g/cm^3
Resistivity	1.59 μΩ-cm
Specific heat capacity	65.9 J/mol. K
Solubility	Soluble in acid, ethanol

3.6.1 XRD Analysis of Silver Oxide Nanoparticles

Each crystalline substance possesses a distinctive atomic arrangement, resulting in a characteristic XRD pattern. These patterns serve as unique identifiers for crystal structures.

A tetragonal type with a measured lattice constant of 6.62 and 5.90 Å (Fig. 3.11). X-ray diffraction, a non-destructive method, is employed to ascertain the crystal structure and purity of nanoparticles[27]. The X-ray diffraction patterns of silver oxide, as documented in this study, reveal prominent peaks at 2θ values ranging from 37.8° to 81.6° of Ag_2O, and their corresponding peaks were indexed as shown in Fig. 3.11. The diffraction peaks match those of the distinctive Ag_2O NPs and demonstrate. The peak intensity, position, and width were analysed through XRD patterns [28]. The Full Width at Half Maximum (FWHM) data was utilized with Scherer's formula to determine the average particle size. These findings have been validated using PANalytical X'Pert software. The average grain size of the Ag_2O nanoparticles generated in this process can be estimated using the Debye–Scherer equation (refer to Eq. 3.8). According to various studies, the estimated mean particle size falls within the range of 4–6.2 nm [27].

To determine the size of nanoparticles, we need to consider certain parameters. These include the mean diameter of the nanoparticles, the wavelength of the X-ray radiation source, and the angular FWHM of the XRD peak at the diffraction angle θ. Based on XRD analysis, the estimated mean size of the nanoparticles is 13.794 nm. Comparing the observed diffraction pattern to a database of known patterns enables the identification of the phase or phases present in the sample. From the observations obtained from XRD, presence of Ag_2O can be confirmed [29].

$$d = \frac{0.9\lambda}{\beta \cos\theta} \tag{3.8}$$

where d, λ, β, and θ are the crystallite size, X-ray wavelength, full width at half maximum (FWHM) of the peak, and Bragg angle, respectively.

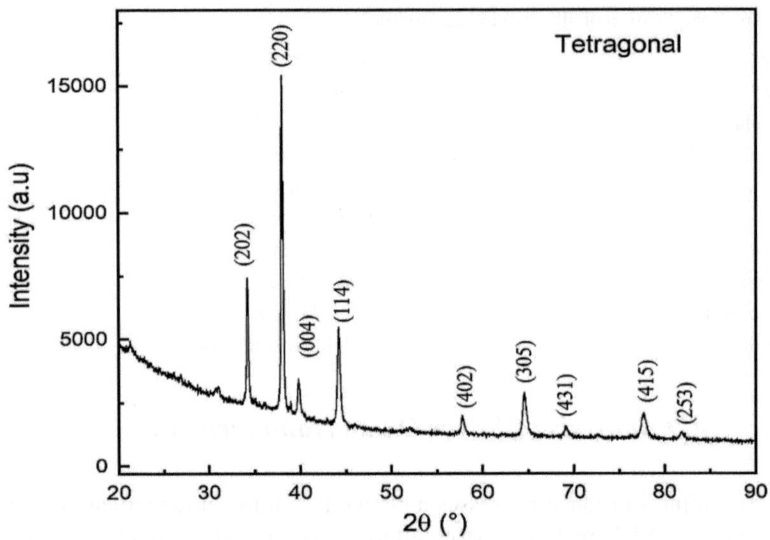

Fig. 3.11 XRD Pattern of Ag_2O NPs

3.6.2 *SEM and EDS of Silver Oxide (Ag_2O) NPs*

The prepared sample was examined via scanning electron microscopy, revealing the morphology of Ag_2O NPs in different resolutions through Fig. 3.12a–d. The images clearly showed that the particles were highly agglomerated and randomly distributed in a spherical shape, with a very high density of growth (Fig. 3.12). The average particle size was measured to be 85.356 nm [30].

3.6.3 *Energy-Dispersive X-Ray Analysis of Ag_2O-NPs*

After analysing the prepared samples using energy-dispersive X-ray analysis, it was found that the samples contained both oxygen and silver elements. The EDS analysis of the sample is shown in Figs. 3.13 and 3.14, which clearly indicate that the sample contained only silver (87.3 wt.%) and oxygen (12.7 wt.%), demonstrating its purity [31]. The XRD analysis also did not show any peaks related to impurities. For further details, please refer to Table 3.7, which provides the elemental analysis of the Ag_2O-NPs sample.

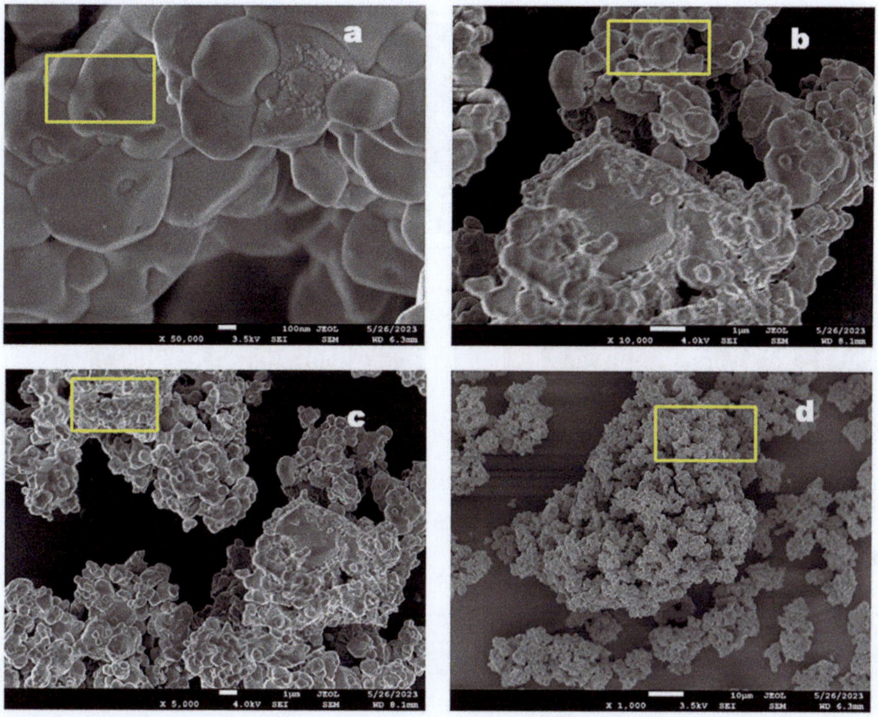

Fig. 3.12 Scanning electron micrographs (SEM) of silver oxide (Ag_2O) NPs

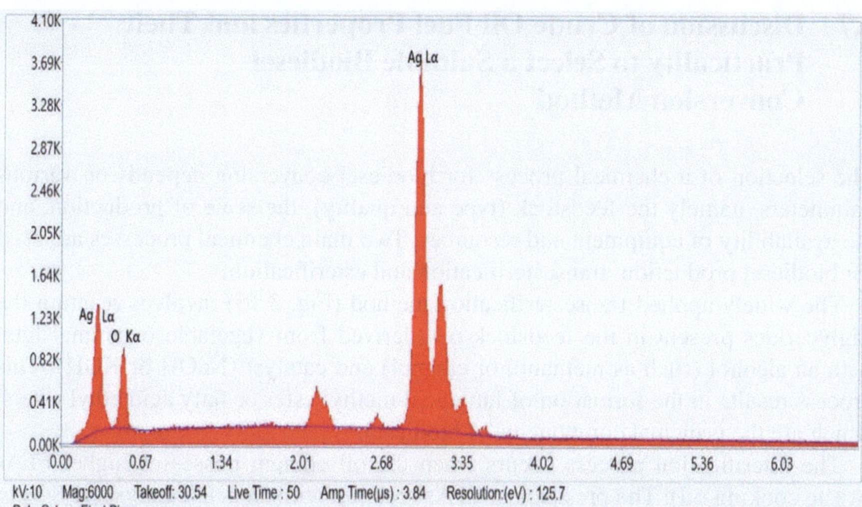

Fig. 3.13 EDS spectrum for Ag_2O NPs

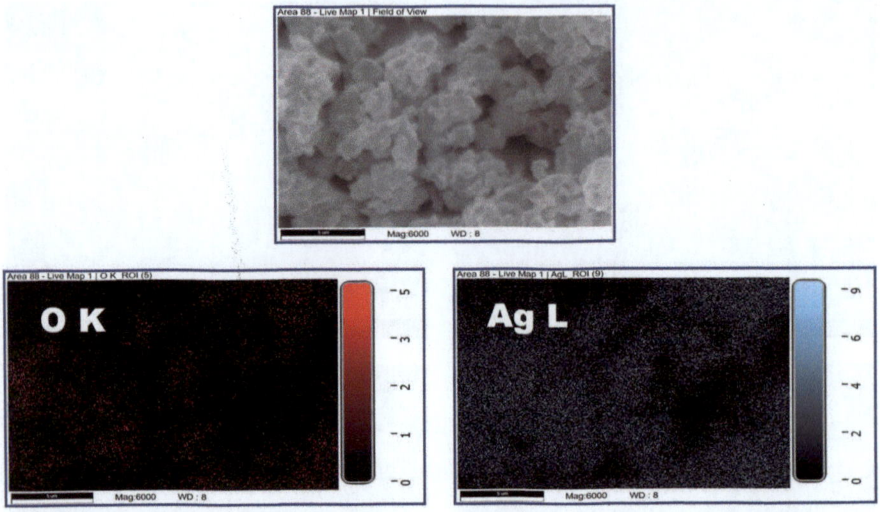

Fig. 3.14 EDS mapping of Ag_2O NPs

Table 3.7 EDS analysis of Ag_2O NPs

Elements	Weight (%)	Atomic (%)
O K	12.7	20
Ag L	87.3	80
Total	100	100

3.7 Discussion of Crude Oil Fuel Properties and Their Practicality to Select a Suitable Biodiesel Conversion Method

The selection of a chemical process for biodiesel conversion depends on various parameters, namely the feedstock (type and quality), the scale of production, and the availability of equipment and resources. Two main chemical processes are used for biodiesel production: transesterification and esterification.

The widely applied transesterification method (Fig. 3.15) involves reacting the triglycerides present in the feedstock oil (derived from vegetable or animal fats) with an alcohol (such as methanol or ethanol) and catalyst (NaOH or KOH). This process results in the formation of fatty acid methyl ester or fatty acid ethyl esters, which are the principal components of biodiesel.

The esterification process occurs when the oil content possesses higher FFAs (waste cooking oil). The presence of FFAs reacting with an acid catalyst (H_2SO_4 or HCl) undergoes a reaction to form esters. The transesterification process is preferred for biodiesel production due to its better efficiency in producing higher yields than esterification.

Fig. 3.15 Transesterification process

However, if the feedstock contains high levels of FFAs, esterification may be a more suitable process. Single-stage transesterification is preferred for FFA <4%, whereas two-stage transesterification is for higher FFA >4%.

3.7.1 Factors to Consider for Biodiesel Production Using Conventional Methods

1. **Quality of raw oil (feedstock)**
   - Ensure that the extracted seed oil is free from sediments, unadulterated, water, and clear.
   - The raw oil should be filtered through 100 μm filter.
2. **Quality of seed**
   - Keep moisture levels at a maximum of 3–5%, sand ensure good, ripe, dried, and properly stored seeds are used. Avoid fungus-attacked seeds.
3. **Quality of methanol**
   - Use clear methanol without any unwanted particles.
   - The density of methanol should exceed 0.79.

4. **Quality of alkali**
   - Use pure sodium hydroxide without sediments. Store NaOH in a clean, dry place away from direct sunlight.
5. **Quality of water for biodiesel washing**
   - Wash biodiesel with clean distilled water at a temperature of 40 °C.

3.8 Procedure for the Transesterification Process

The FFA level for GGG seed oil is more than 4%, indicating that the three-stage process (esterification process) is most suitable for higher yield.

3.8.1 Esterification Process (Three-Stage Process)

Stage 1: The esterification process shown in Fig. 3.16.

1. Take a metered quantity of seed (e.g. 1 L) in a 3-neck flask, place on a magnetic stirrer at a suitable speed (RPM) which is placed in a magnetic pallet and maintains a suitable speed (i.e. RPM). Later, fix the reflux condenser, ensure continuous

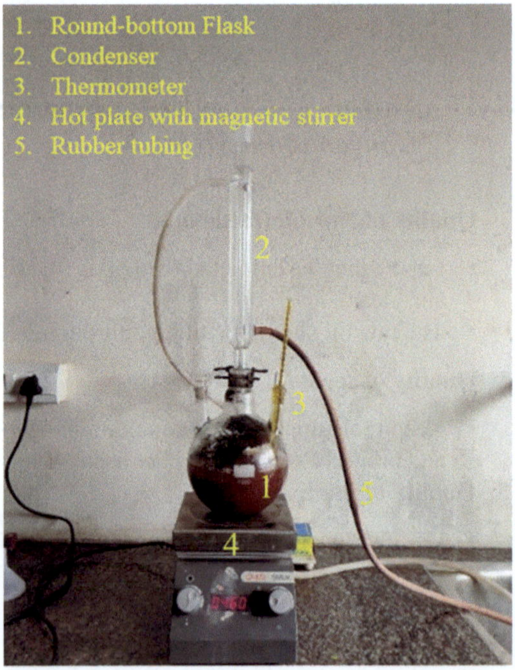

Fig. 3.16 Esterification set-up

Fig. 3.17 Separation process of acid and FFA layers

water circulation, and keep the temperature at 60 °C. Attach a reflux condenser, ensure continuous water circulation, and maintain the temperature at 60 °C.

2. The quantity of H_2SO_4 (for each FFA multiplied by 0.25) and 150 mL of methanol are added slowly to 3-neck flasks. Agitate the prepared solution for 1–1.5 h at a temperature of 60 °C. The said procedure resulted in a dark layer appearing at the top layer of oil (the acid layer appears black) due to a chemical reaction (refer to Fig. 3.17).
3. The mixture is then transferred to another funnel and allowed to least 3 h.
4. The top and bottom layer are separated and stored separately.
5. FFA is determined by taking an oil sample from the 3-neck flask.
6. If the FFA is more than 4%, adjust the RPM to 200 and set the temperature to 60 °C, return to step no-2 and repeat the process until the FFA level is reduced to <4%.

If the FFA is less than 4%, proceed to stage 2.

Stage 2: Production of Sodium Methoxide

1. The metered quantity of 150 mL of methanol per litre of oil and the calculated amount of NaOH based on FFA are mixed thoroughly in a beaker containing the methoxide solution. The above mixture is added slowly to a 3-neck flask to the solution under agitation at 60 °C.
2. The mixture is agitated at a different speed for 60–90 min.
3. Follow the procedure as per stage 3.

Stage 3: Treatment with Base Catalyst

1. A metered quantity of oil (let's say 1 L) is taken in a 3-neck flask and placed on a magnetic stirrer.
2. Insert the magnet pellet inside the flask, and place the reflex condenser on the 3-neck flask, allowing continuous water circulation through the condenser inlet.
3. Turn on the magnetic stirrer connected to electric power and set the suitable operating speed and temperature (60 °C).

4. Insert the thermo-well containing oil with a thermometer in the 3-neck flask to record the temperature.
5. Take the measured quantity (about 300 mL) of methanol/per litre of oil in a 500 mL beaker.
6. Add the measured proportion of NaOH (according to FFA) to the solution containing oil and methanol. Stir the mixture (methoxide) well.
7. Once the pre-set temperature is reached (around 63 °C), slowly add the methoxide mixture to the hot oil in the 3-neck flask, known as the loading part. Maintain the operating speed of the mixture at 400 to 600 RPM, and close the loading port with a stopper.
8. Repeat the process for 30 min while keeping the temperature between 60 and 63 °C. Note the colour change of the mixture to a dull cherry red in the 3-neck flask.
9. Allow the sample to settle in a test tube and ensure separate separation of glycerine and oil occurs.
10. Repeat the procedure every 30 min.
11. Repeat the above steps for a minimum duration of 90 min.
12. Turn off the power supply and transfer the mixture to a separate funnel.
13. After 2 h of settling time, separate the glycerine settled at the bottom and the biodiesel settled at the top layer (refer to Fig. 3.18).

3.8.2 *Recovery of Methanol from Biodiesel*

1. The separated biodiesel or methyl ester is transferred to the reaction vessel of the 3-neck flask.
2. Ensure the set-up has appropriate control over heat, stirring, and distillation. Later, fix the double valve condenser that ensures water circulation and recovery arrangements (refer to Fig. 3.19).

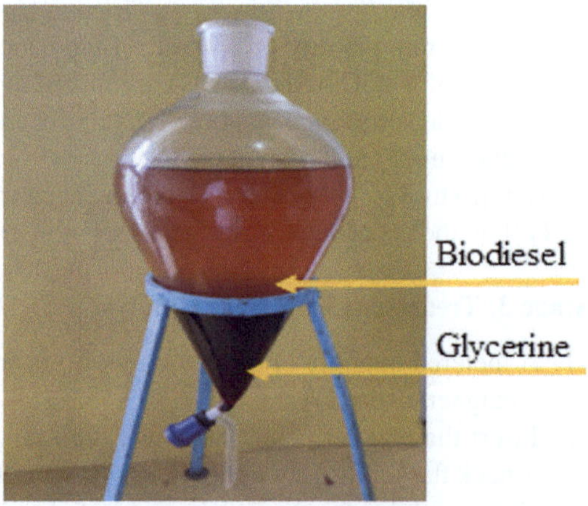

Fig. 3.18 Separation process of biodiesel and glycerine

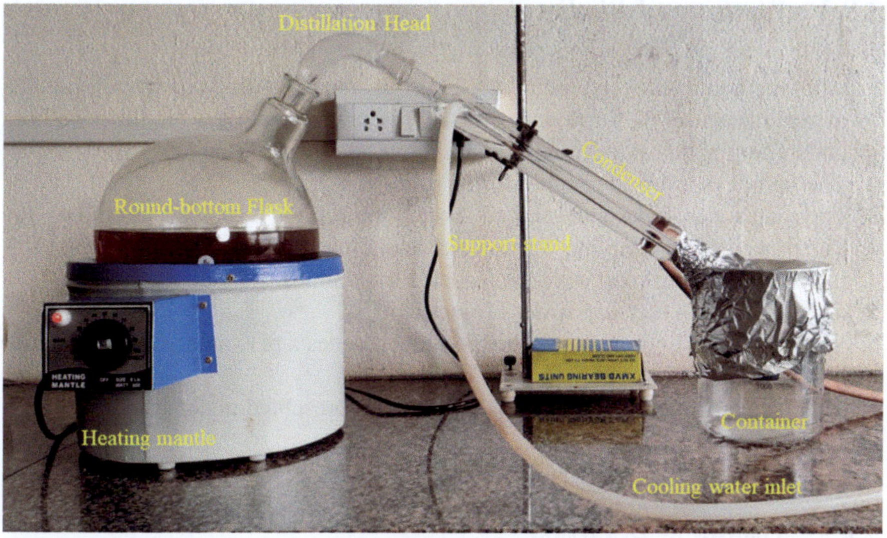

Fig. 3.19 Methanol recovery using condensation process

Fig. 3.20 Recovered methanol

3. Solutions are initiated to agitation at 100 rpm at a temperature of 70 °C.
4. Evaporation of methanol takes place at the pre-set temperature.
5. Record the quantity of methanol collected (refer to Fig. 3.20).
6. Switch off the entire system after ensuring the condensation of methanol stops.

3.8.3 *Washing of Biodiesel*

1. After methanol recovery, the biodiesel is transferred to the wash funnel. Later, a metered quantity of warm water (40 °C) is sprayed into the biodiesel without undergoing agitation (refer to Fig. 3.21).
2. The sample is allowed to settle for 15 min.
3. The soap water formed and settled at the bottom layer is drained carefully.
4. The procedure of spraying warm water and allowing it to settle for 15 min, followed by draining soap water, is repeated three times.
5. The pH value of the soap water is recorded.
6. Water wash is continued until the biodiesel reaches a pH value of seven (refer to Fig. 3.21).
7. The biodiesel extraction from GGG seed oil is presented in detail in Fig. 3.22.

3.8.4 *Drying of Biodiesel*

1. After water wash, the biodiesel is transferred to a 1 L beaker containing a magnetic pellet wherein the set-up is operating at 100 rpm.

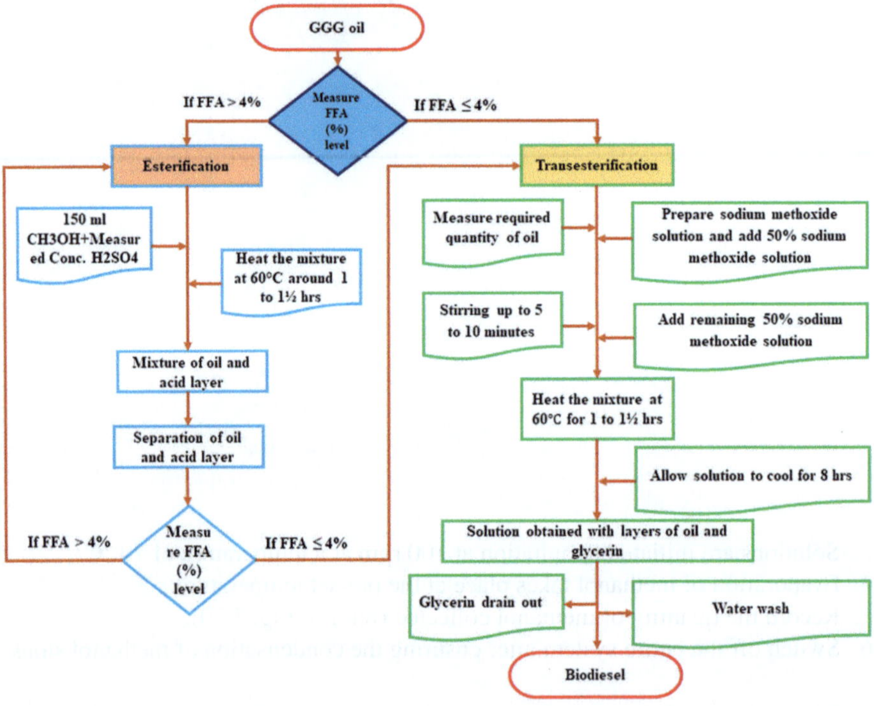

Fig. 3.21 Water wash process

Fig. 3.22 Biodiesel production process

Fig. 3.23 Pure biodiesel obtained

2. The presence of moisture in the biodiesel is removed and subjected to maintaining a temperature of 110–115 °C for 15 min.
3. Allow the biodiesel to cool at room temperature gradually.
4. The quantity of biodiesel obtained after cooling is recorded.
5. Store and clean the biodiesel in a dry container (Fig. 3.23).

3.9 Biodiesel Conversion via Transesterification Process

A mechanical expeller followed by a solvent extraction procedure is employed for oil extraction from the GGG seed-based feedstocks. The literature survey concluded that the proportions of methanol (A): 30, 40, and 50 mL, NaOH (B): 0.6, 0.8, and 1.0 g, and reaction time (C): 45, 60, and 75 min influence biodiesel conversion (yield) [32–34]. A total of 17 experiments (each trial is repeated thrice) with CCD and BBD matrices designed for modelling, and analysis. The average values of biodiesel yield corresponding to each experimental run are computed using Eq. (3.9).

The mass of the biodiesel is calculated based on the sum of the masses of oil, alcohol, glycerol, and unreacted chemicals.

The experimental data corresponding to CCD and BBD matrices were used for parametric analysis. Design Expert Software platform was utilized to develop mathematical modelling and statistical analysis. The following tasks were performed and discussed using Design Expert Software:

1. Parameter analysis: The significance of factors (individual, quadratic, or curvature effect, and interaction among factors) was tested at a 95% confidence level.
2. Two-factor interactions were investigated through the development of three-dimensional surface plots.
3. A mathematical expression for biodiesel yield was derived as a function of transesterification variables.
4. The statistical adequacy of non-linear models was assessed based on correlation coefficient values.
5. Prediction performance analysis was conducted with ten random experimental cases using the derived mathematical expression.

$$\text{Biodiesel yield}(\%) = \frac{\text{Mass of the biodiesel obtained}}{\text{Mass of crude oil taken}} \tag{3.9}$$

3.9.1 *Parametric Analysis*

Two fractional factorial designs (CCD and BBD) were used to develop non-linear models using 17 sets of experimental data. The models derive the mathematical expressions correlating the output and inputs. The mathematical equation (i.e. regression equations) expressing biodiesel yield as a function of transesterification variables based on CCD and BBD experimental data are presented in Eqs. (3.10) and (3.11).

$$\begin{aligned}\text{Yield}_{\text{BBD}} &= 259.7 - 40.75\text{A} + 1053.6\text{B} + 0.963\text{C} + 0.6391\text{A}^2 \\ &\quad - 544.9\text{B}^2 - 0.00232\text{C}^2 - 2.5\text{AB} + 0.0037\text{AC} - 0.717\text{BC}\end{aligned} \tag{3.10}$$

$$\begin{aligned}\text{Yield}_{\text{CCD}} &= 463 - 50.51\text{A} + 1042\text{B} - 0.02\text{C} + 0.7647\text{A}^2 - 620.8\text{B}^2 \\ &+ 0.007\text{C}^2 + 0.31\text{AB} - 0.0124\text{AC} - 0.277\text{BC}\end{aligned} \quad (3.11)$$

3.9.2 *Transesterification Variables Effect Analysis on Biodiesel Yield*

The mean values at each factor level that are affected by main factors (methanol, NaOH, and reaction time) on the performance of biodiesel yield. Methanol showed the highest contribution compared to NaOH and reaction time. A negligible effect (i.e. reaction time between 45 and 75 min does not show maximum changes in biodiesel yield) with reaction time is recorded. A lesser reaction time enables industries to maximize biodiesel production. An increase in methanol quantity tends to increase the biodiesel yield due to increased collision frequency between reactant molecules, resulting in reduced reaction time. Pilot experiments (i.e. >40 mL of methanol) showed increased biodiesel separation from crude oil complexity. The results correlate better with the published literature [35]. In addition, the by-product (glycerol) solubility tends to increase in biodiesel, hindering the transesterification process due to a reversible reaction. The NaOH concentration maintained at the middle levels resulted in the highest biodiesel yield. A lower proportion of NaOH concentration requires more time to attain equilibrium conditions due to the short time to undergo a chemical reaction. However, beyond the critical limit (>0.8 g) of NaOH, resulting in poor diffusion between the molecules (methanol, oil, and catalyst), causes a lower yield. In addition, excessive use of catalysts is not economical and environmentally friendly.

3.9.3 *Analysis of Variance for Biodiesel Yield*

The non-linear models developed based on CCD and BBD were used to test the parametric analysis. Factor analysis (i.e. analysis conducted to determine the parametric significance set at a pre-set 95% confidence level) was carried out, and the model's statistical acceptability was tested for the collected experimental data. Table 3.8 presents the ANOVA results for yield. For both non-linear models, the effect of linear factors (all individual factors effect, i.e. A, B, C) was found to be significant (p-values <0.05). The square terms of methanol and sodium hydroxide were found to be significant, indicating their non-linear which depicts their relationship with biodiesel yield (Table 3.8).

Insignificant contributions with linear variations were observed for reaction time with biodiesel yield, and therefore the p-values of the square term of reaction time were found to be >0.05. The interaction factors effect for the CCD model was

Table 3.8 ANOVA results for the biodiesel yield

Model		BBD				CCD			
Source	DF	Adj. SS	Adj. MS	*F*-value	*P*-value	Adj. SS	Adj. MS	*F*-value	*P*-value
Model	9	4889.7	543.3	194.8	0	4404.6	489.4	034.1	0
Linear	3	1913.7	637.9	228.7	0	2385.7	795.3	055.3	0
Square	3	2932.2	977.4	350.5	0	2005.5	668.5	046.6	0
Interaction	3	43.8	14.6	5.2	0.033	13.4	4.5	0.3	0.817
Error	7	19.5	2.8			100.6	14.4		
Lack of fit	3	8.3	2.8	0.98	0.487	90.9	18.2	3.8	0.224
Pure error	4	11.3	2.8			9.71	4.9		
Total	16	4909.2				4505.2			

Adj. MS adjusted mean squares, *Adj. SS* adjusted sum of squares, *ANOVA* analysis of variance; *BBD* box-behnken design, *CCD* central composite design, *DF* degrees of freedom, *F-value* ratio of two mean squares, *P-value* probability value

insignificant (p-values of two-term interaction >0.05) for biodiesel yield, whereas it was found to be significant (p-values of two-term interaction <0.05) for BBD (refer to Table 3.8). Although both models used identical three factor levels for experimentation, they resulted in different significant and insignificant terms towards biodiesel yield. The probable reason for different significant and insignificant terms is that the points used for experiments are different, leading to different coefficients, p-values, and F-values. The model's statistical acceptability was validated by determining the coefficient of determination (R^2 value). An R^2 value close to 0 depicts no correlation, whereas an R^2 value close to 1 depicts a strong correlation. A higher R^2 value of 0.996 was recorded for BBD than 0.9777 for CCD. Although a higher R^2 value was recorded for BBD, both models showed a better coefficient of determination value, close to unity. Therefore, BBD model data were used to perform the response surface analysis. This indicates that both non-linear models are statistically adequate. Thus, the models can be further validated by conducting prediction and optimization tasks.

3.9.4 *Testing Model Predictions*

The models exhibited a better coefficient of determination, and their practical utility (in terms of generalized models) was assessed through ten random experiments. The percentage deviation values followed a same trend, appearing on both the positive and negative sides of the reference zero line for both models, with a deviation range varying between −6.18 and +1.51 for BBD and −7.21 and +8.45 for

CCD. Additionally, the average absolute percentage deviation corresponding to CCD and BBD is 5.24 and 3.02, respectively. Therefore, the BBD model can be utilized to achieve better prediction accuracy.

3.9.5 Optimization for Maximum Biodiesel Yield

The optimization task could help determine the transesterification parameter values responsible for the maximum biodiesel yield. The BBD-derived response equation is used as an objective function to determine the optimal points (methanol: 40 mL, NaOH 0.83 g, and reaction time: 75 min) responsible for maximum biodiesel conversion. TLBO is applied to perform the optimization task, specifically MATLAB 14. The algorithm runs at least ten times to ensure reliable results, and the average yield was recorded at 93.8%. The highest biodiesel yield of 94.01% is converged to the population size and iteration value equal to 30 and 16, respectively. Experiments were performed for the TLBO to determine optimal conditions resulting in a maximum biodiesel conversion rate equal to 96.2%.

3.9.6 Testing the Optimized Transesterification Factors with Different Weight Proportion of Ag_2O Nano-Catalyst on Biodiesel Yield

The previous section discussed the details of optimized transesterification variables (methanol: 40 mL, NaOH: 0.83 g, and reaction time: 75 min) that resulted in a maximum biodiesel yield of 96.2% experimentally with the determined optimal conditions by the TLBO algorithm. Literature review confirmed the use of nano-catalysts resulting in better performance in biodiesel conversion. Therefore, the NaOH catalyst was replaced with Ag_2O nano-catalyst under the same optimal conditions to assess the practical utility of performance with Ag_2O nano-catalyst towards biodiesel conversions. Experiments were conducted with three replicates for the determined optimal condition (methanol: 40 mL, Ag_2O: 0.83 g, and reaction time: 75 min) and the biodiesel conversion was recorded. The resulting biodiesel yield turned out to be 97.8 ± 0.5%. The Ag_2O nano-catalyst was recovered by filtration followed by washing with methanol and drying at 90 °C. The recovered nano-catalyst was reused for the next cycle of transesterification reaction. It is noteworthy that the biodiesel conversion after six cycles of reuse yielded 90.1% biodiesel conversion (refer to Fig. 3.24). This phenomenon results lowering the biodiesel cost.

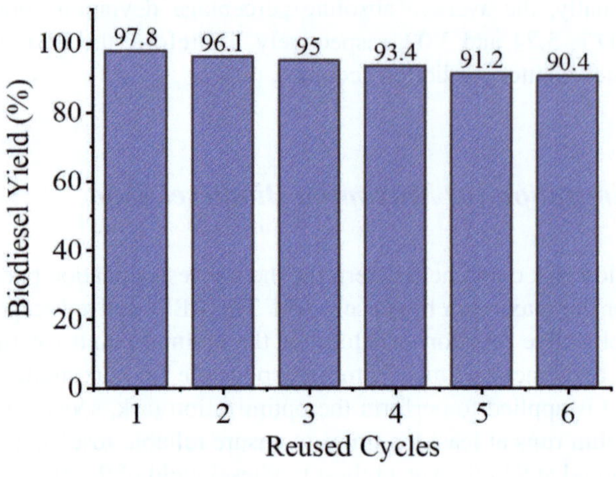

Fig. 3.24 Reuse of Ag_2O nano-catalyst

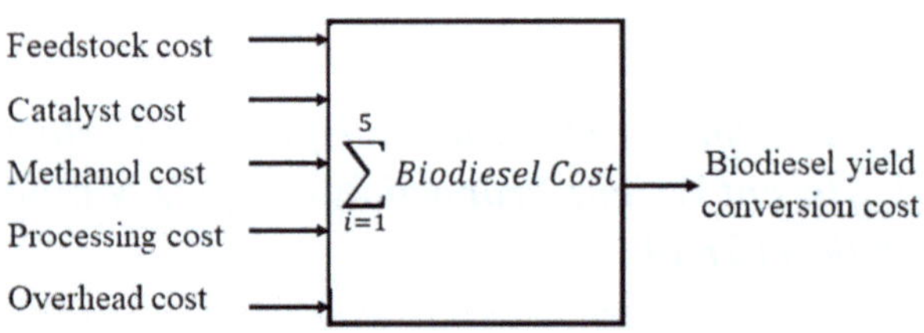

Fig. 3.25 Schematic view of cost estimation Garcinia Gummi-Gutta biodiesel production

3.9.7 *The Cost Analysis Computation for 1 L of Biodiesel Production*

The biodiesel yield conversion from Garcinia Gummi-Gutta-based crude oil is estimated by the proposed method by Das et al. [36]. The biodiesel production cost is computed comprising the price of a litre of Garcinia Gummi-Gutta-based crude oil, methanol, NaOH catalyst, energy consumed (power), process time, and other overhead costs (including labour hours, equipment depreciation, maintenance and repair, insurance, and administrative expenses). The mathematical computation incurred towards biodiesel production cost from Garcinia Gummi-Gutta-based crude oil (refer to Fig. 3.25).

The practical requirements associated with biodiesel production cost to help industries scale up in large volumes are presented in Table 3.9. The costs associated with individual items (feedstock, catalyst dosage, methanol, processing, and overhead) and the composite value of production cost per kg of biodiesel is \$0.8393. It

Table 3.9 Computation of biodiesel production cost

Step	Description	Amount (in USD)	Amount (in INR)
Feedstock	Oil cost for 1 kg of biodiesel production = (quantity × GGG oil cost per kg) = (1 × \$0.60)	\$0.60	₹50.03
Catalyst dosage consumed for 1 kg of biodiesel	0.0083 kg	\$0.033	₹2.75
Methanol required for 1 kg of biodiesel	0.04 kg	\$0.058	₹4.84
Processing cost	Transesterification time (h) × units consumed × cost per unit = (1.25 h × 0.8 × \$0.072)	\$0.072	₹6.00
Net cost (feedstock + catalyst + methanol + processing)	(\$0.60 + \$0.033 + \$0.058 + \$0.072)	\$0.763	₹63.62
Overhead cost (10% of net cost)	10% of \$0.763	\$0.0763	₹6.36
Biodiesel production cost per kg	Cost (net + overhead) = 0.763 + 0.0763	\$0.8393	₹69.98

was inferred from the analysis that the feedstock alone accounts for 72% of the total cost. The catalyst accounts for a significant share, followed by methanol, processing, and overhead costs. The cost can be further reduced substantially by scaling up from laboratory to industrial-scale production in the following ways:

- Bulk purchasing reduces the overall cost of feedstock and reactive materials.
- Advanced technologies such as large reactors can reduce methanol consumption, and better catalysts improve the overall conversion efficiency, lowering production costs.
- The use of advanced process control systems and the implementation of automation further reduce operational costs (minimize labour hours and energy consumption) and human error, ensuring better biodiesel product quality.
- The Garcinia Gummi-Gutta is a miscellaneous species tree wherein the by-products (glycerine), and parts of the tree have commercial applications (grown in non-arable forest land, preparing curries, matchsticks, and furniture, poses Vitamin C and hydroxy citric acid to cure heart diseases, polishing gold and silver, varnish, paints, and so on) which will reduce the overall production cost. The commercial value during the biodiesel extraction process, such as seed cake obtained during oil extraction, can be used as manure or preparing bio-composites, and glycerine obtained during biodiesel purification is not considered during the estimation of biodiesel production cost at the laboratory scale.
- Government policy and regulations help industry personnel with incentives, subsidies, taxes, and so on, ensuring additional revenue systems to offset production costs.

3.9.8 Testing the Biodiesel Yield for Fuel Properties and Check Their Practical Usefulness in Diesel Engines

Therefore, biodiesel blends are prepared with varying percentages of biodiesel in diesel. Three combinations are designed by varying different proportions of diesel and biodiesel with appropriate stirring on a magnetic stirring plate.

The practical usefulness of biodiesel yield requires testing fuel properties, followed by examining performance and emission characteristics when fuelled in diesel engines. India's national policy on biofuels recommends using a lower percentage of biofuel in diesel engines. Therefore, biodiesel blends are prepared with varying percentages of biodiesel in diesel. Three combinations are designed by varying different proportions of diesel and biodiesel with appropriate stirring on a magnetic stirring plate. Three biofuel blended samples, namely B10 (10% biodiesel and 90% of diesel), B20 (20% biodiesel and 80% diesel), and B30 (30% biodiesel and 70% of diesel) were prepared for the present work. The prepared samples were examined for fuel properties and compared with neat diesel, biodiesel, and crude oil (refer to Table 3.10).

The physical properties (density and viscosity) of biodiesel increase with biodiesel content. The B100 fuel sample resulted in the highest physical properties, whereas the lowest was observed with neat diesel. The crude GGG oil (GGGCO) resulted in higher physical properties (possibly due to its higher molecular weight and complex chemical structure) and 17.27% lower calorific value than neat diesel.

Table 3.10 Fuel properties of crude oil, diesel, GGG-derived biodiesel and their blend

Properties	Testing procedure ASTM	GGGCO	Diesel	B10	B20	B30	B100
Viscosity at 40 °C (mm^2/s)	D445	48	2.54	2.87	3.21	3.63	6.12
Density at 15 °C (kg/m^3)	D93	882	815	820	826	831	857
Flash point (°C)	D4052	246	54	66	81	93	183
Higher calorific value (MJ/kg)	D240	38.80	43.82	43.12	42.96	42.65	40.28
Cetane number	D613	58	46	47	48	50	55
Copper corrosion 50 °C, 3 h	D130	1a	1a	1a	1a	1a	1a

Additionally, GGG-derived biodiesel showed comparatively better calorific or heating value and physical properties than biodiesel and its blends. Biodiesel (B100) and its blends (B10, B20, B30) resulted in comparatively higher flash and fire points than diesel, making them easier to handle, transport, and storage.

3.9.9 Error Analysis

The prepared fuel blends were examined to determine their practical usefulness in industries that require estimating uncertainty analysis or error in experiments. Uncertainty error occur in experimental results due to experimental conditions, the associated environment, calibration of instruments and their precision level, observations and recording of the output response. Experiments were repeated three, corresponding to each trial, and the resulting average response values were used to conduct accurate analysis. The instrument accuracy corresponding to each factor from the test results is presented in Table 3.11. The overall uncertainty (OU) corresponding to the experiments is computed using Eq. (3.12), resulting in ±3.15%. The experimental uncertainty error is <5%, ensuring that the results are statistically significant.

$$\mathrm{OU} = \sqrt{(\mathrm{EL})^2 + (\mathrm{FFR})^2 + (\mathrm{CO})^2 + (\mathrm{BTE})^2 + (\mathrm{NO}_X)^2 + (\mathrm{HC})^2 + (\mathrm{BSFC})^2} \quad (3.12)$$

$$\mathrm{OU} = \sqrt{(0.52)^2 + (0.32)^2 + (1.34)^2 + (0.63)^2 + (1.62)^2 + (2.12)^2 + (0.51)^2}$$
$$= \pm 3.15$$

Table 3.11 Summary of results from uncertainty analysis

Factors	Accuracy	Uncertainty (%)
EL (kg)	± 0.1	± 0.52
FFR (L/h)	± 0.02	± 0.32
BSFC (g/kW h)	± 0.04	± 0.51
UHC (ppm)	± 1 ppm	± 2.12
BTE (%)	± 0.4	± 0.63
CO (vol. %)	± 0.01	± 1.34
NOx (ppm)	± 1	± 1.62

3.10 Procedure for Engine Performance, Combustion, and Emission Characteristics

IC engines fuelled with biodiesel and diesel blends are examined for their performance, combustion, and emission characteristics to meet that support meeting the stringent demand for alternative and renewable energy sources (refer to Fig. 3.26).

3.10.1 Performance Characteristics

Following procedure employed to analyse the fuel performance in diesel engines, namely BP, BSFC, and BTE. Figure 3.26 shows the schematic view of the experimental set-up and specifications of the diesel engine, respectively (refer to Table 3.12)

1. The first step is the procedure deployed to test the engine using a dynamometer. The dynamometer allows measuring engine power, torque, and fuel consumption under different operating conditions (refer to Fig. 3.26).

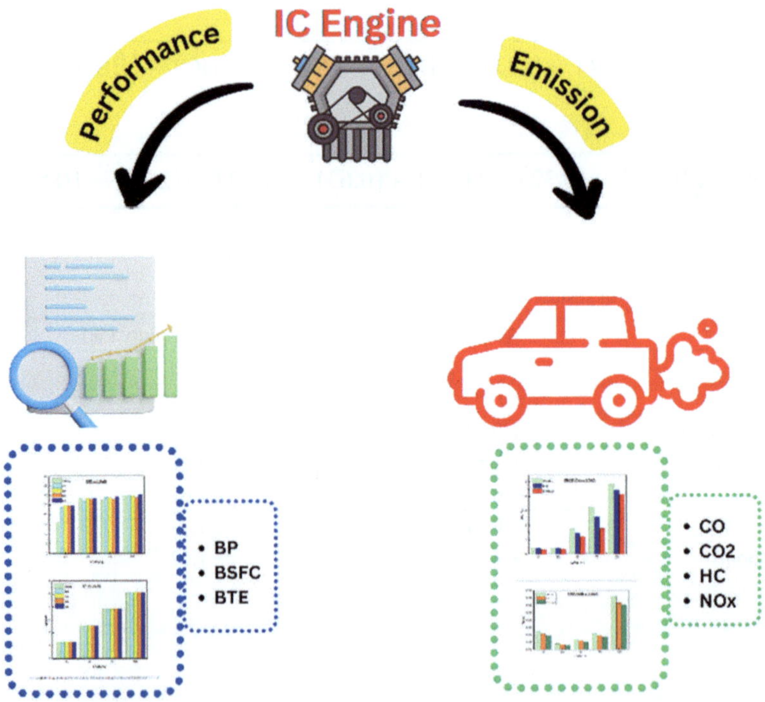

Fig. 3.26 Framework proposed to examine the diesel and biodiesel blends fuelled in diesel engines

Table 3.12 Diesel engine specifications

Engine build	Kirloskar TV1
No. of stroke	4
Fuel	Diesel
No. of cylinder	1
Stroke length	110 mm
Connecting rod length	234 mm
Compression ratio	9.2:1–18:1
Max power	5.20 kW @ 1500 rpm
Cylinder bore	87.50 mm
Orifice diameter	40 mm
Dynamometer arm length	210 mm

2. Engine start-up: Ensure the engine warms up to the predefined operating temperature. The engine should be idled (run the engine at no load) for at least 10 min to ensure a steady-state condition.
3. Perform a baseline test: Conduct a baseline test to establish the engine's power output under specific conditions. Record the engine speed, torque, and fuel consumption.
4. Adjust the load: Adjust the engine load by increasing the dynamometer's resistance. This will increase the torque required to turn the engine resulting in a higher power output.
5. Conduct the test: Increase the load on the engine gradually and record the engine speed, torque, and fuel consumption at each load point.
6. Calculate the BP: The BP measured in kW is calculated using Eq. (3.13).

$$\mathrm{BP} = 2 \times \Pi \times \underbrace{\text{engine speed}}_{\text{rpm}} \times \underbrace{\text{engine torque}}_{N-m} \tag{3.13}$$

7. Calculate the BSFC: The BSFC measured in kg/kWh is computed using Eq. (3.14).

$$\mathrm{BSFC} = \frac{\text{Mass of fuel consumed / unit time}}{\mathrm{BP}} = \frac{\text{kg}}{\text{kw.h}} \tag{3.14}$$

8. Calculate the BTE: The BTE measured in percentage is computed using Eq. (3.15).

$$\mathrm{BTE} = \frac{\mathrm{BP}}{\left(\begin{array}{c}\text{Mass of fuel consumed / unit time} \times \\ \text{calorific value of fuel}\end{array}\right)} = 3.15\% \tag{3.15}$$

3.10.2 Combustion Characteristics

The procedure to find the combustion characteristics in a diesel engine involves the following steps:

1. Install a pressure transducer: Install a pressure transducer in the engine's combustion chamber. The transducer will measure the cylinder pressure during combustion.
2. Calibrate the pressure transducer: Calibrate the pressure transducer to ensure it accurately measures the cylinder pressure.
3. Record the engine speed: Record the engine speed using a tachometer.
4. Record the cylinder pressure: Start the engine and record the pressure at different crank angles during the combustion cycle.
5. Analyse the data: Analyse the data to determine the combustion characteristics of the engine, such as the ignition delay, combustion duration, and combustion efficiency.
6. Repeat the test: Repeat the test at different engine operating conditions, such as other loads or speeds, to determine how the combustion characteristics change under different conditions.

3.10.3 Emission Characteristics

The procedure to find the emission characteristics of an IC engine involves the following steps:

1. Install an exhaust gas analyser: Install an exhaust gas analyser in the engine's exhaust system. The analyser will measure the concentrations of various pollutants in the exhaust gases.
2. Calibrate the exhaust gas analyser: Calibrate the exhaust gas analyser to ensure that it accurately measures the concentrations of pollutants.
3. Record the engine speed and load: Record the engine speed and load using a tachometer and a dynamometer, respectively.
4. Run the engine: Start the engine and run it at a steady state under the desired operating conditions.
5. Collect exhaust gas samples: Collect exhaust gas samples using the exhaust gas analyser.
6. Analyse the exhaust gas samples: Analyse the exhaust gas samples to determine the concentrations of pollutants, such as CO, CO_2, NOx, and HC.
7. Repeat the test: Repeat the test at different engine operating conditions.

3.11 Experimental Analysis and Optimization of GGG-Biodiesel and Its Blends

Figure 3.27 displays the experimental set-up. The technical specification of the engine used for experimentation is presented in Table 3.12. During the experiment, the engine speed and injection pressure are maintained at 1500 rpm and

1. Diesel Engine; 2. Fuel tank; 3. Air filter; 4. Exhaust pipe; 5. Cooling system; 6. Transmission system; 7. Engine mount and base; 8. Gas analyzer; 9. Computer to record readings

Fig. 3.27 Experimental set-up for conducting performance, combustion, and emission tests

Table 3.13 Operating variables and levels for evaluating engine performance and emission characteristics fuelled with GGG-biodiesel

Inputs	Notation	Levels
Engine load (%)	A or EL	20, 60, 100
Injection pressure (bar)	B or IP	160, 210, 260
Compression ratio	C or CR	14, 16, 18
Blend type (vol. %)	D or BT	10, 20, 30

23°bTDC. According to the data, four variables (EL, BT, IP, and CR) have significant effect on engine performance and emission characteristics. These levels were determined after reviewing data from research laboratories and experimental studies [37–40] (refer to Table 3.13).

Before conducting experiments, the engine condition was monitored to ensure stable or steady-state working conditions such as attaining approximately constant temperatures of oil and water. Experiments were conducted with different sets of input variables (EL, IP, CR, and BT) as per the matrices of CCD (refer to Table 3.14). Each experimental run was repeated three times, and the average values of five responses (CO, BTE, BSFC, NOx, and UHC) were used for analysis (refer to Table 3.14). The response-wise study was conducted to understand the insights of influencing variables fuelled with GGG-derived biodiesel.

Table 3.14 provides the details of input-output data corresponding to GGG-derived biodiesel. The data in Table 3.14 were used to derive regression equations for the responses were presented in Eqs. (3.16), (3.17), (3.18), (3.19), and (3.20).

ANOVA helps the investigator to statistically examine the effects of variables that influence the response variables (BSFC, BTE, CO, UHC, and NOx) for the collected experimental input-output data (refer to Table 3.14). The CCD-based non-linear model statistically evaluates the factor effects (all terms of linear or main, square or curvature, and two-factor interaction) for testing significance at a pre-set confidence level of 95% (i.e. p-value <0.05). Design Expert software is employed to

Table 3.14 CCD matrices of GGG-derived biodiesel

Input variables				Output variables				
EL (%)	IP (bar)	CR	BT (vol. %)	BTE, (%)	BSFC (kg/kWh)	CO (vol. %)	UHC (ppm)	NOx (ppm)
100	260	18	10	33.74	0.218	0.27	52	381
20	210	16	20	22.77	0.412	0.14	26	441
100	260	14	30	28.36	0.387	0.21	37	1098
20	160	18	10	24.76	0.331	0.24	43	254
100	160	18	30	30.64	0.378	0.17	37	1084
100	160	18	10	33.18	0.236	0.29	54	334
60	210	18	20	26.5	0.334	0.18	34	678
60	210	16	20	25.78	0.362	0.25	33	564
60	260	16	20	26.18	0.357	0.19	35	616
20	160	18	30	21.63	0.447	0.12	22	1034
100	160	14	30	27.66	0.403	0.2	38	1120
100	260	14	10	32.96	0.252	0.39	67	336
60	210	16	10	25.88	0.283	0.38	61	317
60	210	16	30	23.85	0.412	0.21	32	1073
20	260	14	10	24.14	0.378	0.26	51	227
60	210	16	20	25.94	0.348	0.22	38	594
60	160	16	20	25.13	0.377	0.22	40	567
20	260	14	30	19.56	0.466	0.13	25	943
60	210	14	20	26.14	0.398	0.23	41	512
60	210	16	20	25.74	0.354	0.26	36	588
20	260	18	30	22.43	0.436	0.11	18	1189
20	160	14	30	18.83	0.488	0.15	28	891
100	210	16	20	32.15	0.296	0.24	41	782
20	160	14	10	23.47	0.402	0.31	54	221
20	260	18	10	25.13	0.307	0.22	39	271
100	160	14	10	32.62	0.274	0.45	72	323
100	260	18	30	31.42	0.363	0.18	34	1154

perform the said task. The p-values for each factor are determined, and the factor is termed significant if their p-value <0.05, conversely insignificant if the determined p-value >0.05. The coefficient of determination (R^2 value: considering both the significant and insignificant terms) is computed to know the strength of input-output relationships for each response variable. The model's statistical adequacy is evaluated based on the R^2 value. If the determined R^2 value is close to 1, it signifies the model fits the data well with the assumed response, i.e. better prediction; conversely, a determined R^2 value close to 0 represents the model possessing no correlation (producing worse prediction). The model also determines the adjusted R^2 value (the model with only significant terms), and their value also varies in the ranges between 0 and 1. A lack of fit for the models (evaluated based on p-values) exists for certain responses based on the data. A lack of fit for a model can be made significant by excluding insignificant terms from the model. Estimated *F*-values turnout higher than the tabulated *F*-value after excluding the insignificant terms. Although excluding insignificant terms from the model makes the regression equation simpler, it leads to an improper input-output relationship and affects the prediction precision.

$$\begin{aligned}\mathrm{BTE} = {} & 53.41375 - 0.006986\mathrm{A} + 0.027977\mathrm{B} - 4.21374\mathrm{C} - 0.239568\mathrm{D} \\ & + 0.001038\mathrm{A}^2 - 0.000058\mathrm{B}^2 + 0.130046\mathrm{C}^2 - 0.009348\mathrm{D}^2 \\ & - 0.0000059375\mathrm{AB} - 0.000445\mathrm{AC} + 0.000098\mathrm{D} + 0.000044\mathrm{BC} \\ & + 0.000134\mathrm{BD} + 0.025281\mathrm{CD}\end{aligned} \tag{3.16}$$

$$\begin{aligned}\mathrm{BSFC} = {} & 1.41019 - 0.002548\mathrm{A} - 0.001998\mathrm{B} - 0.09001\mathrm{C} + 0.003605\mathrm{D} \\ & - 0.00000229167\mathrm{A}^2 + 0.00000373333\mathrm{B}^2 + 0.002083\mathrm{C}^2 \\ & - 0.000102\mathrm{D}^2 + 0.0000003125\mathrm{AB} + 0.000072\mathrm{AC} + 0.000021\mathrm{AD} \\ & + 0.00001\mathrm{BC} + 0.000003\mathrm{BD} + 0.000294\mathrm{CD}\end{aligned} \tag{3.17}$$

$$\begin{aligned}\mathrm{CO} = {} & -0.331066 + 0.00589\mathrm{A} + 0.001706\mathrm{B} + 0.114789\mathrm{C} - 0.052457\mathrm{D} \\ & - 0.000021\mathrm{A}^2 - 0.00000748148\mathrm{B}^2 - 0.004676\mathrm{C}^2 + 0.000713\mathrm{D}^2 \\ & + 0.00000125\mathrm{AB} - 0.000141\mathrm{AC} - 0.000019\mathrm{AD} + 0.00005\mathrm{BC} \\ & + 0.000018\mathrm{BD} + 0.000875\mathrm{CD}\end{aligned} \tag{3.18}$$

$$\begin{aligned}\mathrm{UHC} = {} & 161.81035 + 0.48066\mathrm{A} - 0.042681\mathrm{B} - 4.7789\mathrm{C} - 6.83002\mathrm{D} \\ & - 0.002477\mathrm{A}^2 + 0.000015\mathrm{B}^2 + 0.009259\mathrm{C}^2 + 0.09037\mathrm{D}^2 \\ & + 0.000094\mathrm{AB} - 0.000781\mathrm{AC} - 0.000781\mathrm{AD} - 0.000625\mathrm{BC} \\ & + 0.000375\mathrm{BD} + 0.121875\mathrm{CD}\end{aligned} \tag{3.19}$$

$$\begin{aligned}\mathrm{NOx} = {} & -403.66824 + 6.57076\mathrm{A} - 0.369917\mathrm{B} + 46.97569\mathrm{C} \\ & - 16.95181\mathrm{D} + 0.005799\mathrm{A}^2 - 0.004289\mathrm{B}^2 - 1.80556\mathrm{C}^2 \\ & + 0.927778\mathrm{D}^2 - 0.003812\mathrm{AB} - 0.304687\mathrm{AC} - 0.000312\mathrm{AD} \\ & + 0.15\mathrm{BC} + 0.0215\mathrm{BD} + 0.8625\mathrm{CD}\end{aligned} \tag{3.20}$$

3.11.1 Response: BTE

BTE determines the engine efficiency where the heat obtained by burning the fuel is transformed into mechanical energy. Individual and interaction effects of factors were studied based on the main effect plot and surface plots. Analysis of variance was used to examine the statistical significance of factors. The statistical adequacy of the models is validated based on the coefficient of determination (R^2 value).

3.11.1.1 Main Effect Plot of BTE

The effect (relative strength) of individual parameters influencing the BTE is explained through a main effect plot. The main effect plot is constructed based on the average response values corresponding to a design's or input variables' individual levels. Figure 3.28 shows the main effect plot for BTE and is drawn based on the experimental input-output values in Table 3.15.

An increase in engine load, injection pressure, and compression ratio increases the BTE. Engine load was dominant, followed by blend type, compression ratio, and injection pressure. The predominant effect of engine load implies that even a small variation in the load results in a substantial impact (i.e. significant variation in BTE) on BTE. Conversely, injection pressure resulted in a negligible effect (i.e. when varied between 160 and 260 bar, injection pressure resulted in negligible changes observed in BTE) on BTE. Increased compression ratio values improve the combustion phenomenon (air temperature rise inside the engine cylinder), resulting in reduced ignition delay, and increased fuel burning rate causes higher BTE [41]. An increase in fuel injection pressure from 160 to 260 bar showed a marginal rise in

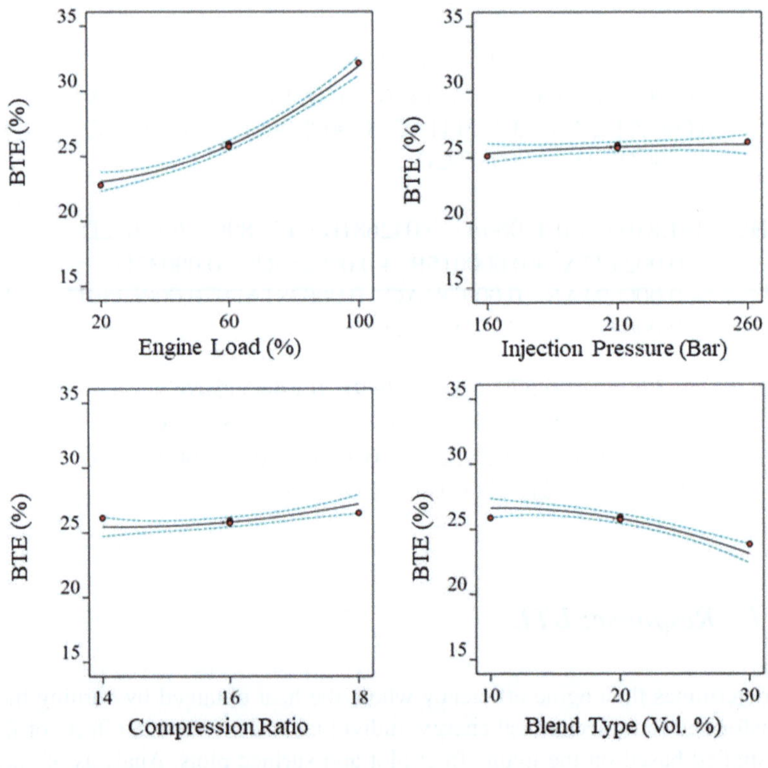

Fig. 3.28 Main effect plot for BTE

Table 3.15 ANOVA results for the BTE: GGG-biodiesel

Source	DF	Adj. SS	Adj. MS	*F*-value	*P*-value
Model	14	443.345	31.668	141.91	0.000
Linear	4	426.446	106.611	477.75	0.000
A: EL	1	355.644	355.644	1593.73	0.000
B: IP	1	2.000	2.000	8.96	0.011
C: CR	1	13.676	13.676	61.29	0.000
D: BT	1	55.125	55.125	247.03	0.000
Square	4	12.690	3.172	14.22	0.000
EL^2	1	7.087	7.087	31.76	0.000
IP^2	1	0.054	0.054	0.24	**0.632**
CR^2	1	0.696	0.696	3.12	**0.103**
BT^2	1	2.247	2.247	10.07	0.008
Interaction	6	4.210	0.702	3.14	0.043
EL × IP	1	0.002	0.002	0.01	**0.922**
EL × CR	1	0.020	0.020	0.09	**0.768**
EL × BT	1	0.025	0.025	0.11	**0.745**
IP × CR	1	0.000	0.000	0.00	**0.971**
IP × BT	1	0.072	0.072	0.32	**0.582**
CR × BT	1	4.091	4.091	18.33	0.001
Error	12	2.678	0.223		
Lack of fit	10	2.655	0.266	23.71	0.041
Pure error	2	0.022	0.011		
Total	26	446.023			

BTE. An increase in injection pressure tends to increase the proportion of smaller or finer fuel droplets entering the combustion chamber because better air-fuel mixture due to proper atomization causes improved combustion efficiency and shortened ignition delay [42]. An engine operating under higher load tends to increase the air temperature, ensuring homogeneous air-fuel mixing results in higher BTE [43, 44]. Increased values of biodiesel content in the prepared blended fuel showed decreased values of BTE (refer to Fig. 3.28). An increase in biodiesel proportion tends to increase the fuel viscosity, density, and latent heat of vaporization, causing inappropriate air-fuel mixing due to the formation of large-size fuel droplets during atomization [45].

3.11.1.2 BTE Surface Analysis

Surface plots are three-dimensional representations of the relationship between the response variable and two input factors varied simultaneously. The following inferences were drawn to examine the two-term interaction effects on BTE:

- Figure 3.29a–c show that BTE efficiency increases with higher values of EL. The combined increase in EL with CR and IP tends to increase BTE (refer to Fig. 3.29a,

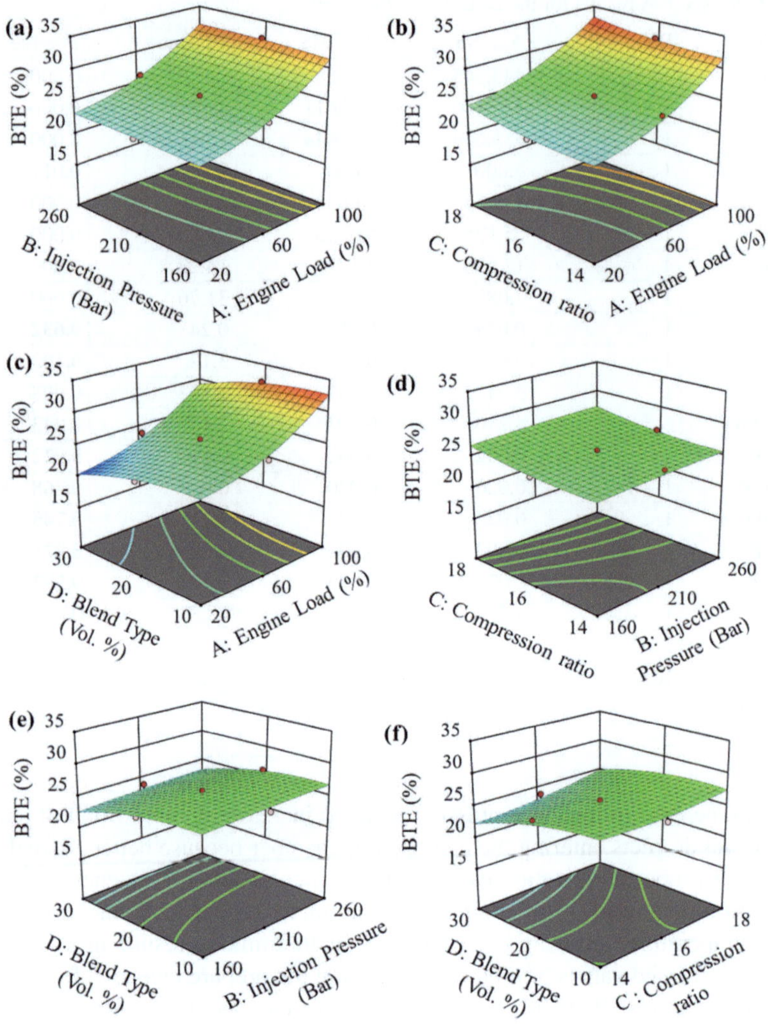

Fig. 3.29 Surface plots of BTE with engine variable: (**a**) EL and IP, (**b**) EL and CR, (**c**) EL and BT, (**d**) IP and CR, (**e**) IP and BT, and (**f**) CR and BT

b). This is because at higher injection pressure, better atomization of the liquid fuel occurs, resulting in smaller droplets entering the combustion chamber with higher jet penetration, leading to improved combustion efficiency. The combined increase in EL and CR results in a rise in air temperature, generating smaller fuel droplets that burn more efficiently. The combined increase in EL and BT also showed increased values of BTE. Increased values of BT resulted in reduced BTE, possibly due to fuel properties (i.e. higher biodiesel content in diesel fuel increases viscosity

and density, leading to poor atomization in the air-fuel mixture). These observations are consistent with literature reports [41, 42, 46].
- The combined increase in IP and CR showed higher BTE (refer to Fig. 3.29d). The results suggest that better fuel atomization with higher IP and combustion efficiency leads higher air temperature inside the engine cylinder with CR [47].
- Figure 3.29e shows the effects of IP and BT on BTE. The effect of IP showed negligible variation compared to BT on BTE. An increased proportion of biodiesel in the blends resulted in a linear decrease in BTE, possibly due to poor atomization and an improper air-fuel mixture [48].
- CR with BT variations on BTE is presented in Fig. 3.29f. An increase in CR from 14 to 18 tends to marginally increase BTE. However, BTE significantly with an increased proportion of biodiesel.

3.11.1.3 ANOVA Analysis of BTE

Table 3.15 presents the ANOVA test results for BTE. The effect of all factors (linear: EL, IP, BT and CR; square: EL^2, IP^2, BT^2, and CR^2; and interaction: EL × IP, EL × CR, EL × BT, IP × CR, IP × BT, CR × BT) are tested for their significance set at a 95% confidence level. Note that the combined effect of all terms in linear, square, and interaction was found to be statistically significant (p-value <0.05). The terms such as EL, IP, BT, CR, EL^2, BT^2, and CR × BT had p-values less than 0.05 indicating their effects on BTE are more pronounced. EL was found to have a dominant effect (when varied within their respective operating ranges), followed by BT and CR for BTE. These effects can be seen in Fig. 3.29 and the F-values in Table 3.15. The determined p-values of square terms such as EL and BT were less than 0.05, indicating a non-linear relationship with BTE (refer to Fig. 3.29). Although all the individual factors were statistically significant, only one interaction term (CR and BT) is significant. This suggests that simultaneous variation in CR and BT results in major changes in BTE (refer to surface plots). The lack of fit is also significant for the BTE model. The model produced better R^2 values and adjusted R^2 values equal to 0.994 and 0.987, indicating the model is statistically adequate and useful for prediction and optimization.

3.11.2 *Response: BSFC*

BSFC determines the fuel efficiency (relies on the air-fuel mixture or rate of fuel consumed) to deliver effective power to the engine. The strength of individual and interaction factor effects was explained by varying within their respective levels and examining the BSFC behaviour with the help of main effect plots and surface plots, respectively (refer to Figs. 3.30 and 3.31).

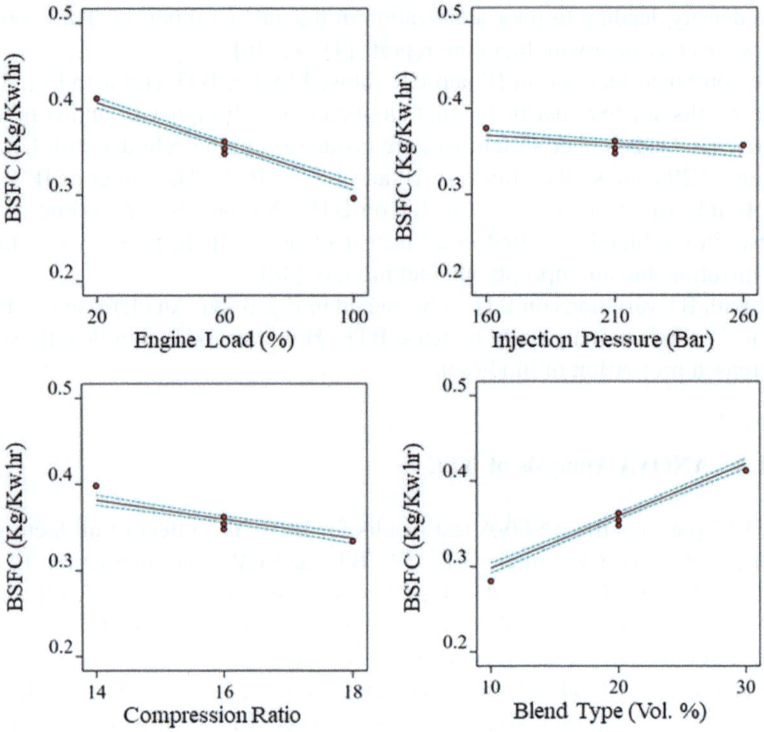

Fig. 3.30 Main effect plot for BSFC

3.11.2.1 Main Effect Plot of BSFC

Figure 3.30 explains the behaviour of BSFC with varying individual variables (EL, IP, BT and CR). BSFC decreases with increased EL (refer to Fig. 3.30). The cylinder pressure increases with a continuous increase in engine load, followed by an increased injected fuel proportion that tends to improve the combustion quality and efficiency, resulting in decreased BSFC [42, 45].

BSFC decreases with increased values of injection pressure, probably due to better atomization and air-fuel mixing processes [49, 50]. BSFC showed a decreasing trend with increased CR (refer to Fig. 3.30). As CR increases, the cylinder pressure increases, causing the fuel to be injected into the hot combustion chamber, generating effective power and, therefore, fuel consumption per unit of power decreases [51]. Increased values of BT (i.e. increased volume proportion of biodiesel in diesel) resulted in increased BSFC. The addition of biodiesel tends to decrease the heating value in the prepared blends, and this phenomenon causes a higher proportion of fuel to be injected into the cylinder to attain the desired power output resulting in increased values of BSFC [47]. BT showed major variations (highest contribution) on BSFC, followed by EL, CR, and IP (refer to Fig. 3.30).

3.11.2.2 Surface Plot Analysis of BSFC

BSFC decreases with a combined increase in IP and EL (refer to Fig. 3.31a). An increase in IP allows for higher jet penetration of fine droplets of fuel to enter the combustion chamber, and higher EL ensures better combustion efficiency and

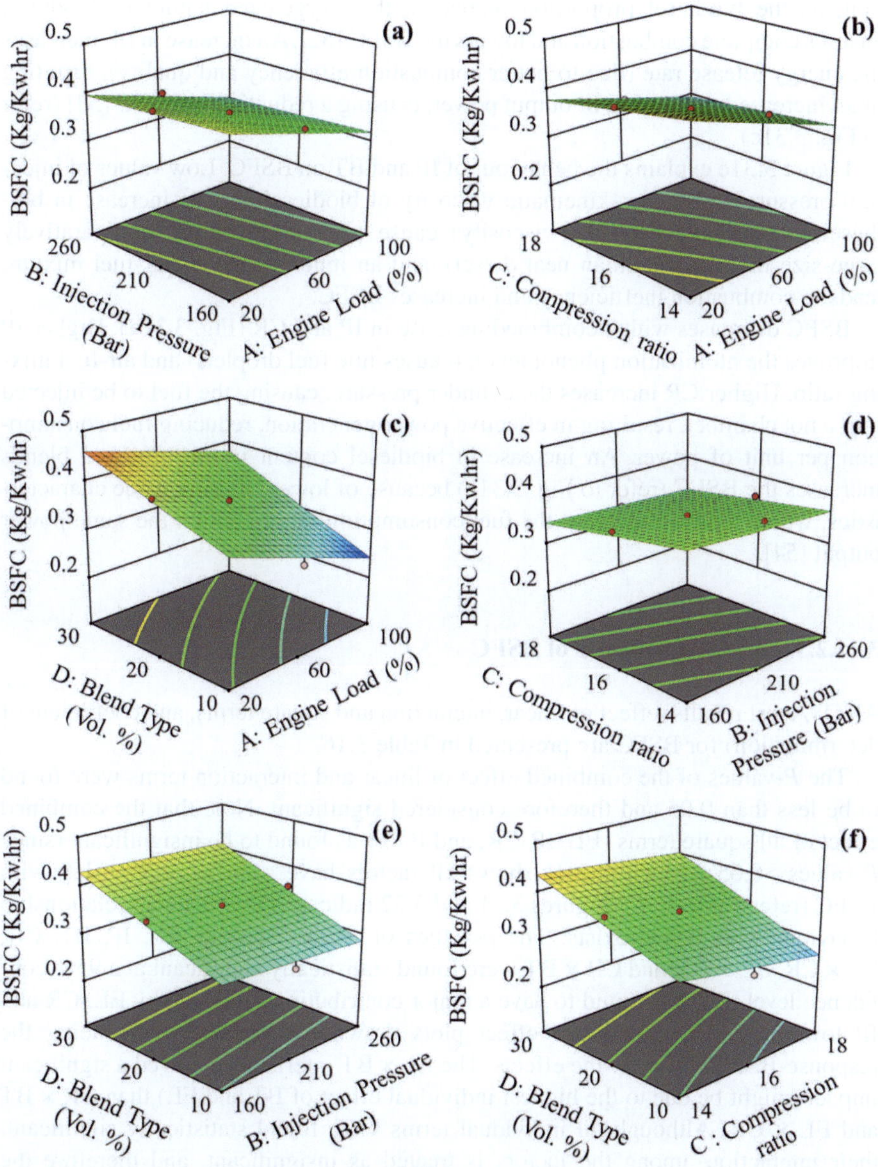

Fig. 3.31 Surface plots of BSFC with engine variable: (**a**) EL and IP, (**b**) EL and CR, (**c**) EL and BT, (**d**) IP and CR, (**e**) IP and BT, and (**f**) CR and BT

quality [49, 50]. Similarly, a higher CR increases the engine cylinder pressure where the fuel is injected into a hot chamber, resulting in better combustion efficiency and generating effective power, leading to reduced BSFC [51] (refer to Fig. 3.31b).

Lower BSFC values were recorded with higher EL and lower BT (lower proportion of biodiesel in neat diesel). An increased proportion of biodiesel in the prepared blends tends to decrease a fuel's calorific value, resulting in lower BSFC. In addition, as the biodiesel proportion increases, the oxygenated nature of biodiesel ensures complete combustion and lowers the BSFC [52]. An increase in EL increases the energy release rate (due to better combustion efficiency and quality), resulting in an increased proportion of output power, causing a reduction in BSFC [53] (refer to Fig. 3.31c).

Figure 3.31e explains the behaviour of IP and BT on BSFC. Low values of injection pressure and higher kinematic viscosity of biodiesel (i.e. an increase in biodiesel content increases the viscosity) cause poor atomization (comparatively large-size fuel droplets than neat diesel), and an inhomogeneous air-fuel mixture leads to combustion inefficiency and increases BSFC.

BSFC decreases with a combined increase in IP and CR (Fig. 3.31d). Higher IP improves the atomization phenomenon (causes fine fuel droplets) and air-fuel mixing ratio. Higher CR increases the cylinder pressure, causing the fuel to be injected into a hot chamber, resulting in effective power generation, reducing fuel consumption per unit of power. An increase in biodiesel content in the prepared blends increases the BSFC (refer to Fig. 3.31e) because of lower calorific value characteristics, which tends to increase the fuel consumption rate to obtain the same power output [54].

3.11.2.3 ANOVA Analysis of BSFC

ANOVA test results (effect of linear, interaction and square terms, and coefficient of determination) for BSFC are presented in Table 3.16.

The *P*-values of the combined effect of linear and interaction terms were found to be less than 0.05 and therefore considered significant. Note that the combined effect of all square terms (EL, IP, CR, and BT) was found to be insignificant (since *P*-values >0.05), which clearly shows all factors have a linear relationship with BSFC (refer Table 3.16). Figures 3.31 and 3.32 indicate a strong linear relationship behaviour with BSFC exists. The *p*-values of factors such as EL, IP, BT, CR, EL × CR, EL × BT, and CR × BT were found statistically significant at a 95% confidence level. BT was found to have a major contribution, followed by EL, CR and IP towards BSFC. The main effect plots drawn for variables influencing the response-BSFC can show the effects. The EL × BT interaction showed a significant impact (might be due to the highest individual effect of BT and EL) than CR × BT and EL × CR. Although all individual terms were found statistically significant, their interaction among the factors is treated as insignificant, and therefore the resulting surface plot is almost flat with negligible variation in BSFC (refer to Fig. 3.30). The lack of fit exists for the model can be made significant by excluding

Table 3.16 ANOVA results for the BSFC: GGG-biodiesel

Source	DF	Adj. SS	Adj. MS	*F*-value	*P*-value
Model	14	0.121467	0.008676	135.93	0.000
Linear	4	0.118633	0.029658	464.65	0.000
A: EL	1	0.041089	0.041089	643.74	0.000
B: IP	1	0.001644	0.001644	25.75	0.000
C: CR	1	0.008800	0.008800	137.87	0.000
D: BT	1	0.067100	0.067100	1051.25	0.000
Square	4	0.000606	0.000151	2.37	0.111
EL^2	1	0.000035	0.000035	0.54	0.476
IP^2	1	0.000224	0.000224	3.51	0.086
CR^2	1	0.000179	0.000179	2.80	0.120
BT^2	1	0.000266	0.000266	4.16	0.064
Interaction	6	0.002228	0.000371	5.82	0.005
EL × IP	1	0.000006	0.000006	0.10	0.760
EL × CR	1	0.000529	0.000529	8.29	0.014
EL × BT	1	0.001089	0.001089	17.06	0.001
IP × CR	1	0.000016	0.000016	0.25	0.626
IP × BT	1	0.000036	0.000036	0.56	0.467
CR × BT	1	0.000552	0.000552	8.65	0.012
Error	12	0.000766	0.000064		
Lack of fit	10	0.000667	0.000067	1.35	0.498
Pure error	2	0.000099	0.000049		
Total	26	0.122233			

the insignificant terms (EL^2, IP^2, BT^2, CR^2, EL × IP, EL × IP, IP × CR, and IP × BT) from the model. Excluding insignificant terms results in estimated *F*-values that are greater than the table *F*-values. In addition, the regression equations become much simpler, which leads to improper input-output relationships with reduced prediction precision. The R^2 and adjusted R^2 value for BSFC are equal to 0.9937 and 0.9864, respectively. Therefore, the model is statistically significant and useful for predicting and optimizing.

3.11.3 Carbon Monoxide Emission

CO emissions are the toxic gases (colourless and odourless) formed due to incomplete combustion resulting from an incomplete air-fuel mixture [55]. The CO emission occurs because the generated carbon atoms might not turn into CO_2, and the settled CO is probably caused by incomplete combustion due to insufficient oxygen. CO emissions are harmful to both the environment and the atmosphere resulting in adverse effects on human health, causing direct poising effects [56]. The main effect plot explains the impact of individual factors, whereas interaction factor effects are

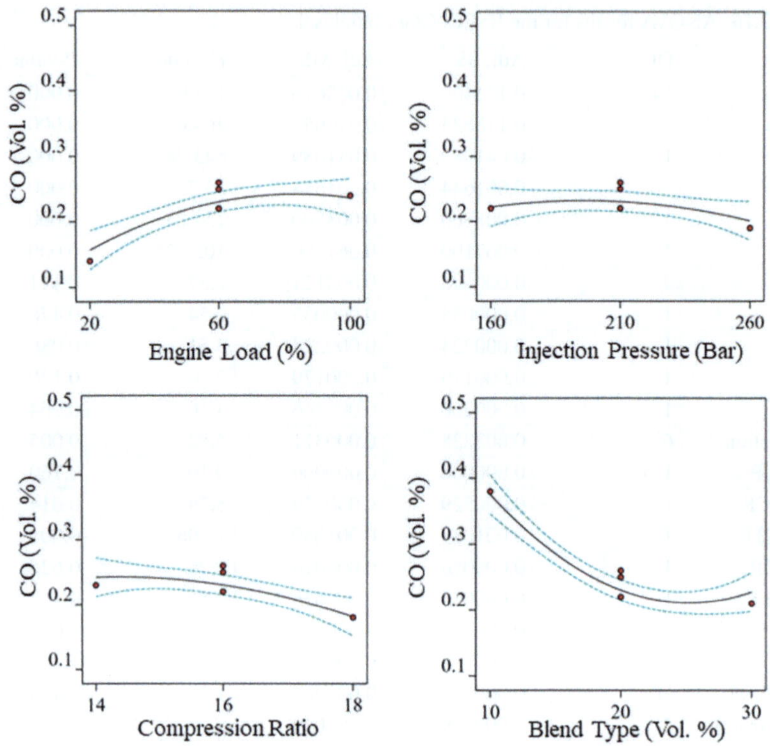

Fig. 3.32 Main effect plot for CO

presented with the help of main effect plots and surface plots (refer to Figs. 3.32 and 3.33). ANOVA table explains the factors' effects and model adequacies for their practical usefulness (refer to Table 3.17).

3.11.3.1 Main Effect Plot of CO Emission

Figure 3.32 explain the behaviour of individual factor on CO emission. Exhaust CO emissions are evaluated at low, medium, and high engine loads. An increase in EL ensures more fuel is injected in a reduced time, resulting in increased formation of CO emissions in diesel engines [57]. CO emissions decreased after crossing the mid-values of EL due to improved combustion characteristics resulting from a better air-fuel mixture [56]. An increase in injection pressure decreases CO emissions. Injection pressure minimizes the dominant effect of the viscosity of blended fuel, and the presence of oxygenated characteristics of biodiesel ensures complete combustion resulting in reduced CO emissions [53]. Low values of CR decrease the CO emission, probably due to inadequate combustion temperature and delayed ignition. At higher CR, the temperature and pressure are closer to the higher top dead centre at the end of compression, and the resulting phenomenon changes the speed of

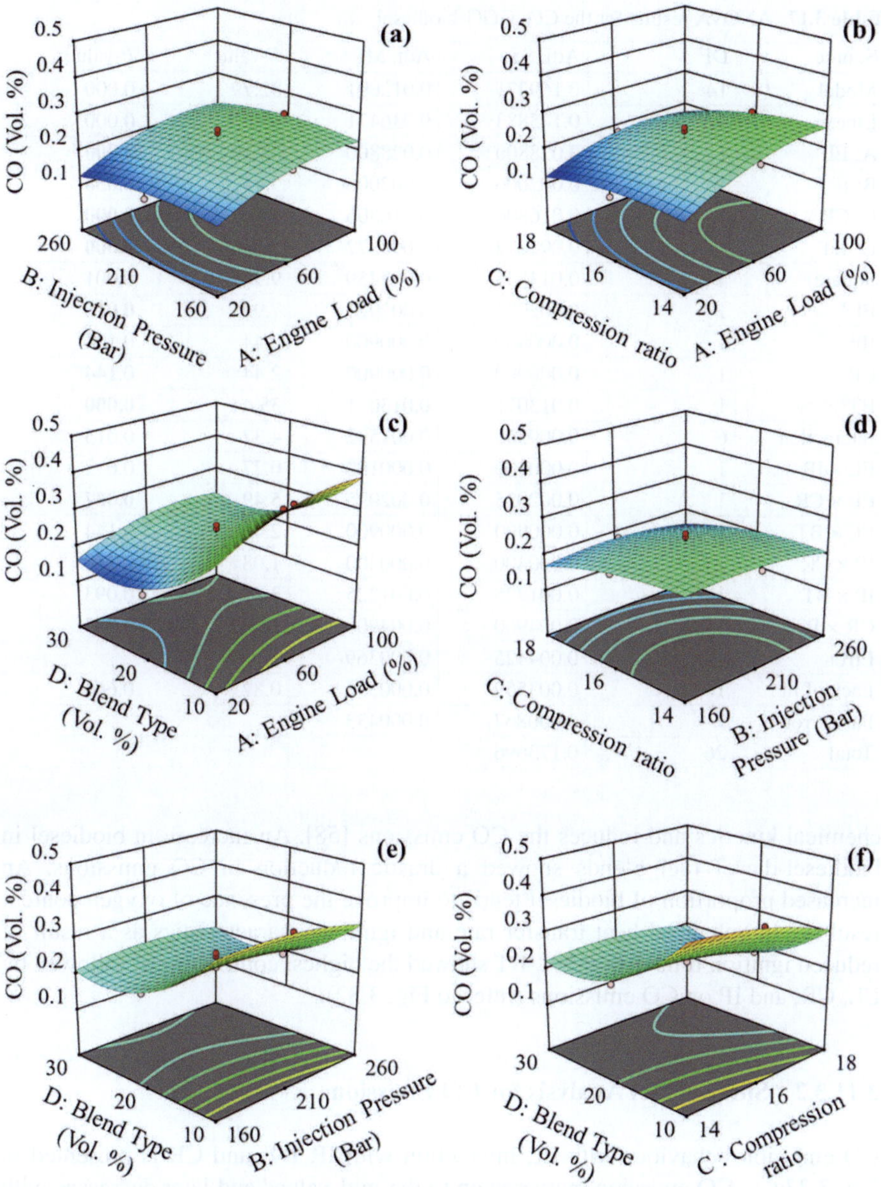

Fig. 3.33 Surface plots of CO emission with engine variable: (**a**) EL and IP, (**b**) EL and CR, (**c**) EL and BT, (**d**) IP and CR, (**e**) IP and BT, and (**f**) CR and BT

Table 3.17 ANOVA results for the CO: GGG-biodiesel

Source	DF	Adj. SS	Adj. MS	*F*-value	*P*-value
Model	14	0.169271	0.012091	32.79	0.000
Linear	4	0.145883	0.036471	98.90	0.000
A: EL	1	0.028800	0.028800	78.10	**0.000**
B: IP	1	0.002006	0.002006	5.44	**0.038**
C: CR	1	0.016806	0.016806	45.57	**0.000**
D: BT	1	0.098272	0.098272	266.48	**0.000**
Square	4	0.013838	0.003459	9.38	0.001
EL^2	1	0.002921	0.002921	7.92	**0.016**
IP^2	1	0.000900	0.000900	2.44	0.144
CR^2	1	0.000900	0.000900	2.44	0.144
BT^2	1	0.013071	0.013071	35.44	**0.000**
Interaction	6	0.009550	0.001592	4.32	0.015
EL × IP	1	0.000100	0.000100	0.27	0.612
EL × CR	1	0.002025	0.002025	5.49	**0.037**
EL × BT	1	0.000900	0.000900	2.44	0.144
IP × CR	1	0.000400	0.000400	1.08	0.318
IP × BT	1	0.001225	0.001225	3.32	0.093
CR × BT	1	0.004900	0.004900	13.29	**0.003**
Error	12	0.004425	0.000369		
Lack of fit	10	0.003559	0.000356	0.82	0.664
Pure error	2	0.000867	0.000433		
Total	26	0.173696			

chemical kinetics and reduces the CO emissions [58]. An increase in biodiesel in biodiesel-diesel fuel blends showed a drastic reduction in CO emissions. An increased proportion of biodiesel tends to improve the presence of oxygen content resulting in enhanced heat transfer rate and ignition characteristics as a result of reduced ignition time delay [59]. BT showed the highest contributions, followed by EL, CR, and IP on CO emissions (refer to Fig. 3.32).

3.11.3.2 Surface Plot Analysis for CO Emission

CO emission behaviour with EL interaction with IP, BT, and CR is presented in Fig. 3.33a–c. CO emission increases up to the mid-values and later decreases with EL. As the engine load increases, more fuel is injected at a reduced time, and if time is insufficient, it causes an increased formation of CO emissions [57]. CO emission decreases after crossing the middle values due to improved combustion characteristics (increased temperature and pressure) [56]. The increased combination of EL with IP, EL with CR, and EL with BT values showed lower CO emissions. The effect of IP interaction with EL, CR, and BT on CO emissions is presented in Fig. 3.33a, c, d. An increase in IP showed a marginal reduction in CO emissions.

This is postulated that fine fuel droplets are created with higher jet pressure caused by increased IP resulting in better combustion efficiency and quality [49, 50]. Combining higher IP values with CR and IP with BT resulted in lower CO emissions. This is attributed to increased temperature and pressure, causing enhanced changes favouring the chemical kinetics of fine droplets of oxygenated fuel injected into a hot chamber, ensuring improved combustion characteristics.

The effect of CR interaction with EL, IP, and BT on CO emissions is presented in Fig. 3.33b, d, f. CR varied from 14 to 18, resulting in decreased values of CO emissions. Low CR delays the ignition due to insufficient heat, whereas an increase in CR improves the air temperature inside the cylinder and thereby reduces the ignition delay, causing the complete combustion of fuel [60]. Higher CR with IP and EL causes fine droplets of fuel to enter the hot chamber caused by higher engine cylinder temperature and pressure at the end of compression set at higher EL, which reduces the ignition delay resulting in complete fuel combustion. The combined high BT and CR values reduced CO emissions (refer to Fig. 3.33f). An increased proportion of biodiesel in biodiesel-diesel blends causes complete combustion due to more oxygen molecules in the fuel supply chain [61].

BT interaction with EL, IP, and CR effects on CO emissions is presented in Fig. 3.33c, e, f. The biodiesel in biodiesel-diesel blend decreases CO emissions. The engine operated at higher IP, EL, and CR improves the atomization and air-fuel mixture, which changes the chemical kinetics resulting in fine droplets of fuel injected into the hot chamber and reduced CO emission [60]. In addition, the increased proportion of oxygenated molecules available in the chemical structure of biodiesel fuel aids in complete combustion, causing reduced CO emission [62].

3.11.3.3 ANOVA Analysis of CO Emission

The ANOVA tests are conducted to examine the parametric significance (linear, interaction, and curvature) of CO emissions presented in Table 3.17. The *P*-values of linear, interaction, and square terms are significant. The effect of EL, IP, CR, and BT was found statistically significant, indicating that these parameters contribute more towards CO emission. BT was found to have a significant contribution, followed by BT, CR, and IP. EL and BT's relationship with CO emissions is non-linear, as their *P*-values were found to be less than 0.05. The *P*-value statistics were found to have good agreement with the main factor and surface plot effects (refer to Table 3.17 and Figs. 3.32 and 3.33). Although all individual factors' influence was found statistically significant, their interaction effects representing EL × IP, EL × BT, IP × CR, IP × BT are found insignificant. Therefore, the resulting surface plots do not show significant variation in CO (refer to Fig. 3.34a, c–e). The CR × BT and EL × CR terms were found to make significant contributions, and their resulting surface plots showed more variations in CO emissions. The CR × BT interaction effect showed a maximum effect on CO emissions and is probably due to the highest individual contributions of CR and BT. Excluding the insignificant terms (IP^2, CR^2, EL × IP, EL × BT, IP × CR, and IP × BT) from the model makes the lack-of-fit term statistically significant. The lack of fit exists for the model can be made significant by excluding the insignificant terms (EL^2, IP^2, BT^2, CR^2, EL × IP, EL × IP, IP × CR, and IP × BT) from the model. Excluding insignificant terms turns out the estimated *F*-values greater than table *F*-values, resulting in an imprecise regression equation and reduced prediction accuracy. The model R^2 (inclusion of both significant and insignificant terms) and adjusted R^2 value were found to be equal to 0.9745 and 0.9448. Therefore, the model is statistically significant to make accurate predictions and can be applied to perform optimization.

3.11.4 Unburnt Hydrocarbon Emissions

UHC emissions result from incomplete combustion (due to an inappropriate air-fuel mixture) of fuel. UHC is mainly influenced by engine parameters, fuel properties (such as the presence of oxygen content), and spray characteristics [63]. The effects of factors, including individual, combined, and curvature effects, are illustrated using main effect and surface plots (see Figs. 3.34 and 3.35). An analysis of variance table is constructed to statistically explain the factors' contributions and model adequacies for practical use (refer to Table 3.18).

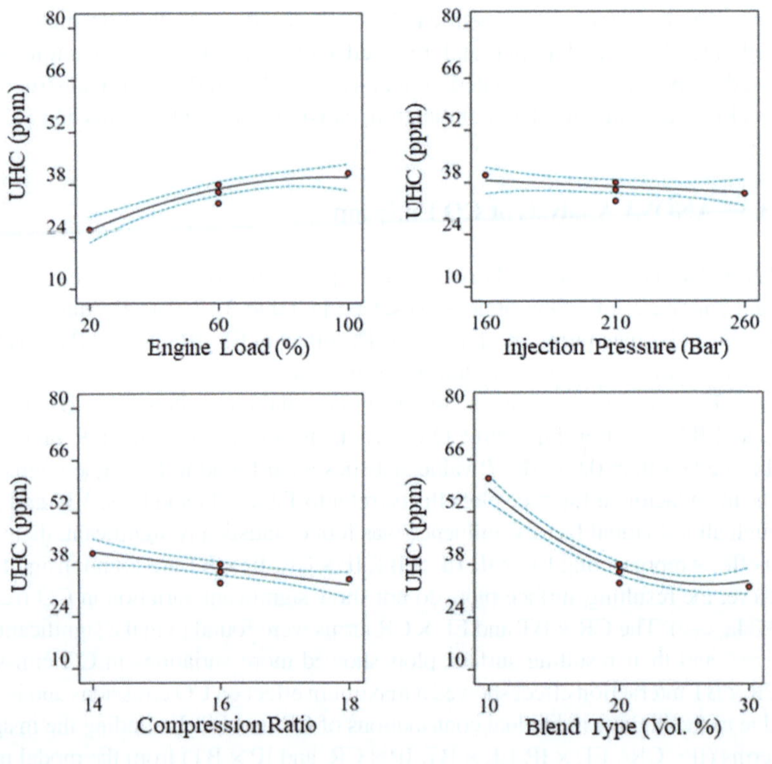

Fig. 3.34 Main effect plot for UHC

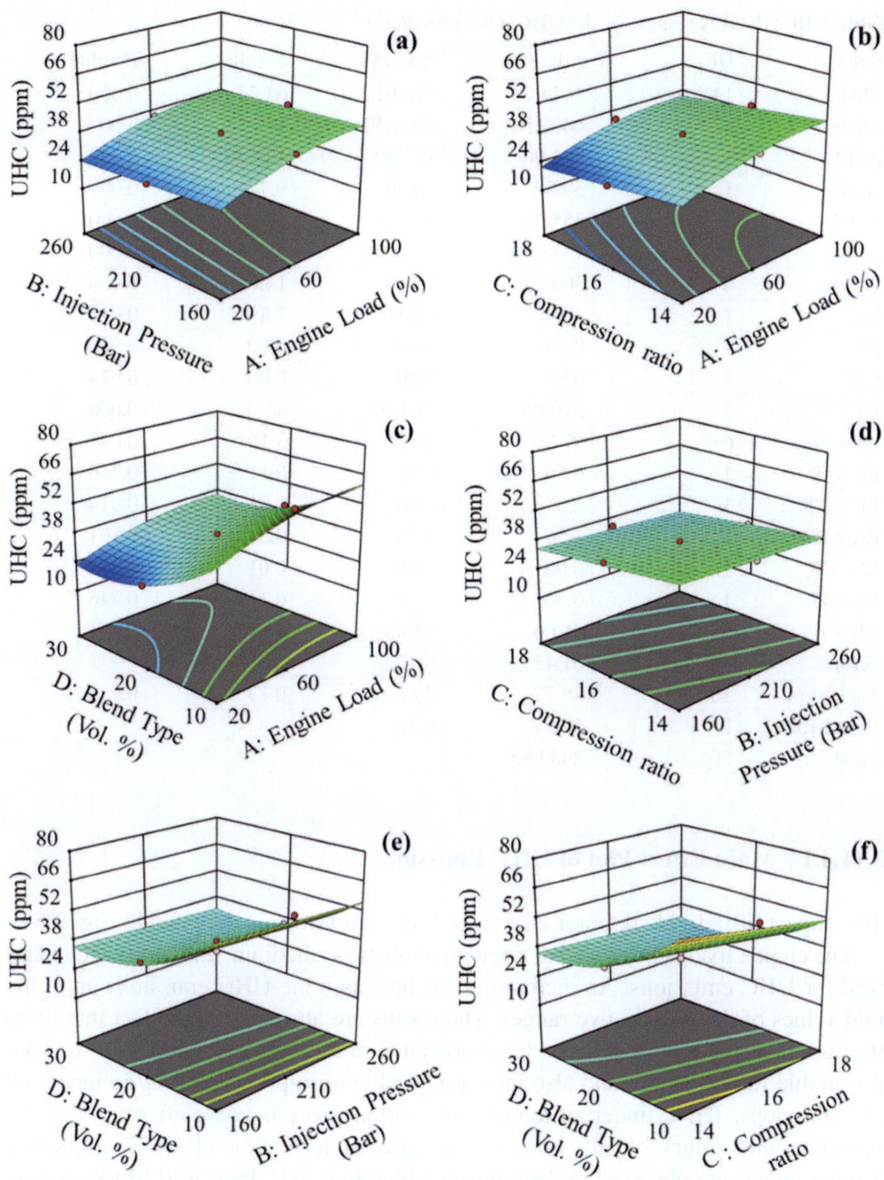

Fig. 3.35 Surface plots of UHC emission with engine variable: (**a**) EL and IP, (**b**) EL and CR, (**c**) EL and BT, (**d**) IP and CR, (**e**) IP and BT, and (**f**) CR and BT

Table 3.18 ANOVA results for the UHC: GGG-biodiesel

Source	DF	Adj. SS	Adj. MS	*F*-value	*P*-value
Model	14	4424.20	316.01	61.73	0.000
Linear	4	4025.56	1006.39	196.58	0.000
A: EL	1	882.00	882.00	172.28	0.000
B: IP	1	50.00	50.00	9.77	0.009
C: CR	1	355.56	355.56	69.45	0.000
D: BT	1	2738.00	2738.00	534.82	0.000
Square	4	300.77	75.19	14.69	0.000
EL^2	1	40.38	40.38	7.89	0.016
IP^2	1	0.00	0.00	0.00	0.979
CR^2	1	0.00	0.00	0.00	0.979
BT^2	1	210.00	210.00	41.02	0.000
Interaction	6	97.88	16.31	3.19	0.041
EL × IP	1	0.56	0.56	0.11	0.746
EL × CR	1	0.06	0.06	0.01	0.914
EL × BT	1	1.56	1.56	0.31	0.591
IP × CR	1	0.06	0.06	0.01	0.914
IP × BT	1	0.56	0.56	0.11	0.746
CR × BT	1	95.06	95.06	18.57	0.001
Error	12	61.43	5.12		
Lack of fit	10	48.77	4.88	0.77	0.685
Pure error	2	12.67	6.33		
Total	26	4485.63			

3.11.4.1 Main Effect Plot of UHC Emission

The impact of individual factor effects on UHC emissions is depicted in Fig. 3.34.

The engine load is increased at three intervals (low, medium, and high) and examined for UHC emissions. An increase in EL increases the UHC emissions up to the mid-values of their respective ranges. The results are attributed to the fact that up to the mid-values of EL, the in-cylinder temperature of the engine might be low, was not favourable for oxidation, and also increased fuel consumption resulting in increased HC emissions. The cylinder air temperature and pressure increase after crossing the operating mid-values of EL, resulting in reduced formation of UHC emissions. Similar effects are observed in the published literature [64]. Increased IP values tend to decrease UHC emissions. Higher fuel injection pressures improve the fuel droplets and ensure fine fuel droplets enter the combustion chamber due to better atomization, resulting in better complete combustion efficiency and reduced UHC [63]. Inadequate heat of compression delays the ignition resulting in higher UHC at low values of CR. However, advancing CR causes improved combustion temperature and pressure resulting in better combustion efficiency and reducing UHC emissions [60]. UHC emissions decrease with increased BT (higher biodiesel percentage in blended diesel-biodiesel fuel), probably due to the higher oxygen content and lower carbon content in biodiesel, ensuring improved combustion quality [64, 65].

3.11.4.2 Surface Plot Analysis of UHC Emission

The combined effect of EL and IP on UHC is presented in Fig. 3.35a. The combination of lower IP and lower-higher EL resulted in reduced UHC emissions. Low fuel consumption at lower EL and superior air-fuel mix due to better atomization phenomenon with higher IP improved the combustion efficiency and shortened the ignition delay resulting in reduced UHC. The combined high values of EL and CR improve the combustion temperature and pressure resulting in improved combustion quality and reduced UHC emission. The combined effect of IP and CR on UHC is presented in Fig. 3.35d. Higher CR and FIP reduced UHC emissions attributed to atomized fine fuel droplets at the nozzle exit, causing more vapour-distributed phases and better homogeneous air-fuel mixes, ensuring complete combustion. Higher CR with enough oxygenated fuel content ensures better combustion quality and thermal efficiency, causing lower UHC in Fig. 3.35 [42]. Increased oxygenated biodiesel proportion ensures uniform air-fuel mixes when the fine fuel droplets exit the nozzle in vapour form with higher IP, causing better combustion quality and, in turn, reduced UHC (refer to Fig. 3.35). The higher proportion of oxygenated biodiesel entering the combustion chamber raises the cylinder air temperature, and pressure increases with higher EL, ensuring complete combustion and reduces UHC.

3.11.4.3 ANOVA Analysis of UHC Emission

The linear factors (El, IP, CR, and BT), interaction factors (El^2, IP^2, CR^2, and BT^2), and quadratic factors (El × IP, EL × CR, EL × BT, IP × CR, IP × BT, and CR × BT), were tested for significance at a 95% confidence level. The terms El, IP, CR, BT, El2, BT2, and CR × BT were found to be statistically significant, with *P*-values less than 0.05. Among these factors, BT had the highest impact (*F*-value of 534.82), followed by EL, CR, and IP on UHC emissions. This implies that even a small change in the percentage of biodiesel in biodiesel-diesel blended fuel can lead to significant changes in UHC emissions. The main effect plot for BT on UHC emissions aligns well with the statistical results (see Fig. 3.35 and Table 3.18). The square terms of IP and CR were found to be insignificant, indicating a linear relationship with UHC emissions. The variations in UHC emissions with changes in factors, both linear and non-linear, are consistent with the statistical results and plots (refer to Table 3.18 and Figs. 3.34 and 3.35). The interaction factor CR × BT was found to be statistically significant, suggesting that an increase in both CR and BT leads to significant changes in UHC emissions. Minimal variations in UHC emissions were observed with simultaneous changes in El × IP, EL × CR, EL × BT, IP × CR, and IP × BT factors. The lack-of-fit model term was found to be insignificant. The correlation coefficient and adjusted R^2 value were calculated to be 0.9863 and 0.9703, respectively. A goodness of fit close to >98.5% indicates that the resulting model accurately reflects the experimental results.

3.11.5 *Nitrogen Oxide Emissions*

NOx emissions result from gases formed when the fuel is burnt in the combustion chamber at higher temperatures [66]. Controlling the engine parameters and fuel characteristics reduces NOx emissions. The main effect plot and surface plots effectively explain the individual and interaction factors. The analysis of the variance table is constructed to explain the statistical influence of factors and the model's practical usefulness (refer to Table 3.19).

3.11.5.1 Main Effect Plot of NOx Emission

Figure 3.36 explains the individual factor effects on NOx emissions. Individually, the factors varied between their respective levels and showed increased NOx emissions. The BT showed a dominant effect accompanied by EL, CR, and IP on NOx (refer to Fig. 3.36). An increase in BT increases the biodiesel percentage in the fuel, which delivers higher oxygen content during combustion resulting in increased exhaust temperature, causing higher NOx [60]. NOx emission increases with an increase in fuel IP, probably due to the increased fuel droplet diameter that causes

Table 3.19 ANOVA results for the NOx: GGG-biodiesel

Source	DF	Adj. SS	Adj. MS	*F*-value	*P*-value
Model	14	2,837,056	202,647	56.42	0.000
Linear	4	2,770,389	692,597	192.82	0.000
A: EL	1	72,327	72,327	20.14	0.001
B: IP	1	8321	8321	2.32	**0.154**
C: CR	1	27,848	27,848	7.75	0.017
D: BT	1	2,661,894	2,661,894	741.09	0.000
Square	4	46,021	11,505	3.20	0.052
EL^2	1	221	221	0.06	**0.808**
IP^2	1	296	296	0.08	**0.779**
CR^2	1	134	134	0.04	**0.850**
BT^2	1	22,134	22,134	6.16	0.029
Interaction	6	20,647	3441	0.96	0.492
EL × IP	1	930	930	0.26	0.620
EL × CR	1	9506	9506	2.65	0.130
EL × BT	1	0	0	0.00	0.993
IP × CR	1	3600	3600	1.00	0.337
IP × BT	1	1849	1849	0.51	0.487
CR × BT	1	4761	4761	1.33	0.272
Error	12	43,103	3592		
Lack of fit	10	42,599	4260	16.90	0.057
Pure error	2	504	252		
Total	26	2,880,159			

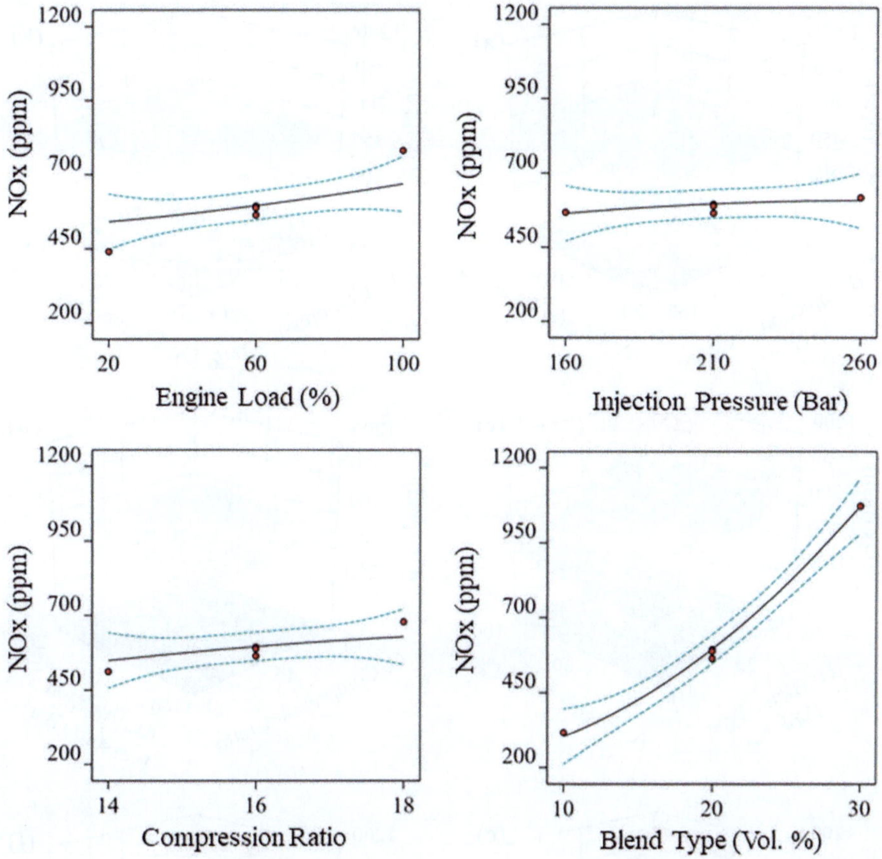

Fig. 3.36 Main effect plot on NOx emission

the fuel spray to vaporize quickly and favours faster combustion rates resulting in higher exhaust temperature [47]. Advancing CR towards a higher value tends to increase the cylinder temperature at the end of the compression stroke resulting in higher NOx [67]. At low EL values, the engine cylinder's temperature is lower, resulting in incomplete combustion. Increases in EL tend to increase fuel consumption and raise the cylinder temperature and pressure, ensuring increased exhaust temperature causes higher NOx emission at high values of EL [67].

3.11.5.2 ANOVA and Surface Plots of NOx Emission

The effect of full quadratic terms such as El, IP, CR, BT, El^2, IP^2, CR^2, BT^2, El × IP, EL × CR, EL × BT, IP × CR, IP × BT, and CR × BT on the engine NOx performance was analysed at 95% confidence level. The effect of IP is found to be insignificant, indicating that the varied levels of IP from 160 to 260 bar resulted in a negligible change in NOx emission (refer to Figs. 3.36 and 3.37). The BT effect on NOx

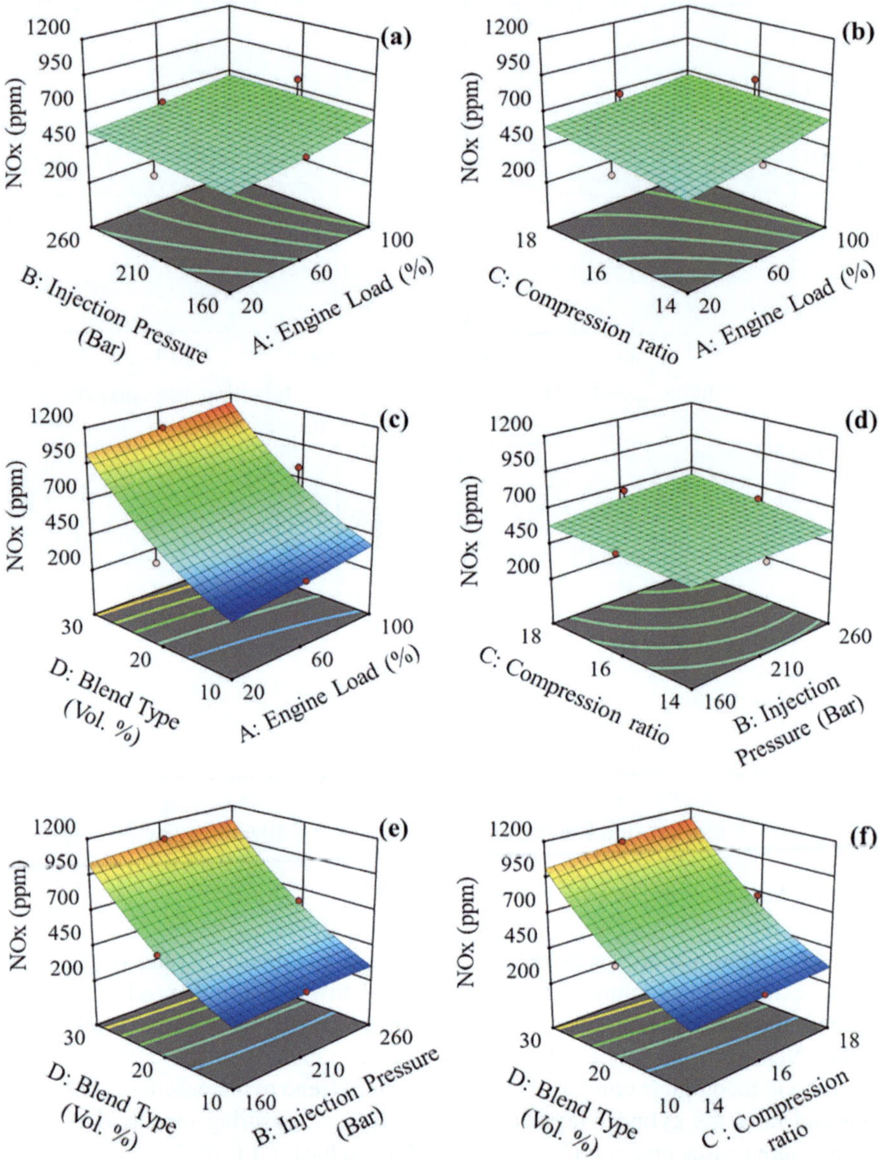

Fig. 3.37 Surface plots of UHC emission with engine variable: (**a**) EL and IP, (**b**) EL and CR, (**c**) EL and BT, (**d**) IP and CR, (**e**) IP and BT, and (**f**) CR and BT

emission showed the highest contribution, followed by EL and CR. Figures 3.36 and 3.37 showed that NOx emissions have linear relationships with EL, IP, and CR, as their square term *P*-values >0.05. All individual terms (excluding IP) showed significant individual contributions. However, none of their interaction effects were found to be insignificant. The resulting surface plot appears almost flat, as shown in Fig. 3.37a–f. Therefore, the resulting surface plots appear almost flat. Including insignificant interaction terms such as El × IP, EL × CR, EL × BT, IP × CR, IP × BT, and CR × BT in regression equations does not show significant changes in NOx emission. The lack-of-fit model term was found to be insignificant at the pre-set 95% confidence level. The R^2 and adjusted R^2 values were equal to 0.9850 and 0.9676. The resulting better R^2 value is close to 1, indicating the model is statistically adequate to make better predictions.

3.12 Concluding Remarks

The experimental set-up, equipment, and different methods used for oil extraction, as well as the procedures for fuel properties and testing standards employed to quantify different responses. The characterization of feedstocks and nanoparticles is also explained. The procedures for performance, combustion, and emission characteristics are outlined.

The GGG seed oil is transformed into biodiesel through the transesterification process. Two non-linear regression models, namely CCD and BBD, were used to provide a comprehensive understanding of the transesterification process. The effect of each transesterification process variable (methanol, NaOH, reaction time) on biodiesel yield was studied using main or individual effects and interaction or surface plots. Methanol had the highest contribution, followed by NaOH and reaction time on biodiesel yield. All interaction terms (excluding NaOH with Methanol) were found to be significant in relation to biodiesel yield. The empirical equation derived for biodiesel yield as a mathematical function of transesterification variables allowed for the prediction of 10 experimental cases with a mean absolute percent deviation of 3.02% for BBD and 5.24% for CCD, respectively. Therefore, a BBD-derived empirical equation was used to determine the optimal transesterification conditions using the TLBO algorithm. TLBO recommended 40 mL of methanol, 0.83 g/100 cc of NaOH, and a reaction time of 75 min as the optimal conditions that maximize predicted biodiesel yield to 94.01%. The experimental biodiesel yield under these conditions was 96.2%. The efficacy of biodiesel yield with silver oxide nanoparticle for optimized condition is found to 97.8% for first cycle and yielded 90.1% after sixth cycles of reuse. The estimated cost of biodiesel conversion accounts to $0.8393.

The tested fuel properties (viscosity, density, flash point, calorific value, cetane number, and copper strip corrosion) of GGG-derived biodiesel and its blends are valuable for practical applications. The overall uncertainty in experiments was determined to be ±3.15% to assess practical utility in industries. Biodiesel and its blends (B10: 10% biodiesel and 90% diesel, B20: 20% biodiesel and 80% diesel, and B30: 30% biodiesel and 70% diesel) were tested for practical use in diesel engines without any modifications. The significance of full quadratic (linear, square,

Table 3.20 Responses for diesel engine features

Output	Linear	Square	Interaction	Coefficient of determination
BTE	EL, CR, IP, BT	EL^2, BT^2	CR × BT	$R^2 = 0.9940$; Adj. $R^2 = 0.9870$
BSFC	EL, CR, IP, BT	EL^2, BT^2, CR^2, IP^2	EL × CR, EL × BT, CR × BT	$R^2 = 0.9937$; Adj. $R^2 = 0.9864$
CO	EL, CR, IP, BT	EL^2, BT^2	EL × CR, CR × BT	$R^2 = 0.9745$; Adj. $R^2 = 0.9448$
UHC	EL, CR, IP, BT	EL^2, BT^2	CR × BT	$R^2 = 0.9863$; Adj. $R^2 = 0.9703$
NOx	EL, CR, BT	EL^2, BT^2, CR^2	NIL	$R^2 = 0.985$; Adj. $R^2 = 0.9676$

and interaction) terms and the statistical adequacies of non-linear models for different responses are summarized in Table 3.20.

All linear terms (except IP for NOx emission) were statistically significant for performance and emission characteristics. The contribution of individual factors according to their importance was listed for different responses EL > BT > CR > IP for BTE, BT > EL > CR > IP for BSFC, and CO, UHC, and NOx emissions. The engine load and blend type were found to vary non-linearly with the performance and emission characteristics. The CR × BT interaction term was significant (except for NOx) towards all responses. All models produced a better coefficient of determination close to 1, indicating that the models are statistically adequate for performing better prediction and perform optimization.

References

1. Raghavendra, M., Yatish, K., Lalithamba, H.: Plant-mediated green synthesis of ZnO nanoparticles using Garcinia gummi-gutta seed extract: photoluminescence, screening of their catalytic activity in antioxidant, formylation and biodiesel production. Eur. Phys. J. Plus. **132**, 1–12 (2017)
2. Nair, K., Mohanan, C., Mathew, G.: Preliminary observations on seed and plantation technology for Garcinia gummi-gutta (Guttiferae): an indigenous and economic tree crop of the Indian Peninsula. For. Trees Livelihoods. **15**(1), 75–87 (2005)
3. Sudharani, N., et al.: Estimation of bioactive components and antioxidant activity of garcinia gummigutta fruit rind. J. Pharmacogn. Phytochem. **7**(3S), 467–470 (2018)
4. HS, Y.K., et al.: Nutraceutical and medicinal values of minor fruits in Western Ghats of South India. J. Pharmacogn. Phytochem. **7**(3S), 404–408 (2018)
5. Ravishankar, K.V., et al.: Isolation and characterization of microsatellite markers in Garcinia gummi-gutta by next-generation sequencing and cross-species amplification. J. Genet. **96**, 213–218 (2017)
6. Semwal, R.B., et al.: A comprehensive scientific overview of Garcinia cambogia. Fitoterapia. **102**, 134–148 (2015)

7. Jena, B., et al.: Chemistry and biochemistry of (−)-hydroxycitric acid from Garcinia. J. Agric. Food Chem. **50**(1), 10–22 (2002)
8. Kumode, M.M.N., et al.: Microfibrillated nanocellulose from balsa tree as potential reinforcement in the preparation of 'green' composites with castor seed cake. J. Clean. Prod. **149**, 1157–1163 (2017)
9. Hegde, M.B., et al.: Garcinia gummigutta vegetable oil–graphene oxide nano-composite: an efficient and eco-friendly material for corrosion prevention of mild steel in saline medium. J. Polym. Environ. **28**, 483–499 (2020)
10. Contran, N., et al.: State-of-the-art of the Jatropha curcas productive chain: from sowing to biodiesel and by-products. Ind. Crop Prod. **42**, 202–215 (2013)
11. Konaka, T., et al.: Use of carbonized fallen leaves of Jatropha curcas L. as a soil conditioner for acidic and undernourished soil. Agronomy. **9**(5), 236 (2019)
12. Lima, P.J.M., et al.: An overview on the conversion of glycerol to value-added industrial products via chemical and biochemical routes. Biotechnol. Appl. Biochem. **69**(6), 2794–2818 (2022)
13. Hejna, A., et al.: Potential applications of crude glycerol in polymer technology–current state and perspectives. Renew. Sustain. Energy Rev. **66**, 449–475 (2016)
14. Hanumantha, M., et al.: Utilization pattern of non-timber forest products (NTFPS) in Siddapura taluk of Uttara Kannada district of western Ghat region, Karnataka. J. Pharmacogn. Phytochem. **8**(3), 901–905 (2019)
15. Atacan, K., et al.: Fabrication of silver doped different metal oxide nanoparticles and evaluation of their antibacterial and catalytic applications. Appl. Surf. Sci. Adv. **6**, 100156 (2021)
16. Nakamura, S., et al.: Synthesis and application of silver nanoparticles (Ag NPs) for the prevention of infection in healthcare workers. Int. J. Mol. Sci. **20**(15), 3620 (2019)
17. Rahman, M.M., et al.: Highly sensitive methanol chemical sensor based on undoped silver oxide nanoparticles prepared by a solution method. Microchim. Acta. **178**, 99–106 (2012)
18. Shetty, V.: Solar light active biogenic titanium dioxide embedded silver oxide (AgO/Ag2O@TiO2) nanocomposite structures for dye degradation by photocatalysis. Mater. Sci. Semicond. Process. **132**, 105923 (2021)
19. Choi, H., et al.: Versatile surface plasmon resonance of carbon-dot-supported silver nanoparticles in polymer optoelectronic devices. Nat. Photonics. **7**(9), 732–738 (2013)
20. Burdușel, A.-C., et al.: Biomedical applications of silver nanoparticles: an up-to-date overview. Nanomaterials. **8**(9), 681 (2018)
21. Gadkari, R.R., et al.: Silver nanoparticles in water purification: opportunities and challenges. Modern age environmental problems their remediation, pp. 229–237 (2018)
22. Kraśniewska, K., Galus, S., Gniewosz, M.: Biopolymers-based materials containing silver nanoparticles as active packaging for food applications–a review. Int. J. Mol. Sci. **21**(3), 698 (2020)
23. Dawood, S., et al.: Biodiesel synthesis from Prunus bokhariensis non-edible seed oil by using green silver oxide nanocatalyst. Chemosphere. **291**, 132780 (2022)
24. Ahmad, T., et al.: Low temperature chemical synthesis and comparative studies of silver oxide nanoparticles. J. Mol. Struct. **1084**, 9–15 (2015)
25. Siddiqui, M.R.H., et al.: Synthesis and characterization of silver oxide and silver chloride nanoparticles with high thermal stability. Asian J. Chem. **25**(6), 3405–3409 (2013)
26. Ocsoy, I., et al.: A green approach for formation of silver nanoparticles on magnetic graphene oxide and highly effective antimicrobial activity and reusability. J. Mol. Liq. **227**, 147–152 (2017)
27. Thiruvengadam, V., Bansod, A.V.: Characterization of silver nanoparticles synthesized using chemical method and its antibacterial property. Biointerface Res. Appl. Chem. **10**, 7257–7264 (2020)
28. Shume, W.M., Murthy, H.A., Zereffa, E.A.: A review on synthesis and characterization of Ag2O nanoparticles for photocatalytic applications. J. Chem. **2020**(1), 5039479 (2020)
29. Rezaee Jordehi, A.: A chaotic-based big bang–big crunch algorithm for solving global optimisation problems. Neural Comput. Applic. **25**, 1329–1335 (2014)

30. Zhang, Y., et al.: Effects of nitrogen content on the structure and mechanical properties of (Al0. 5CrFeNiTi0. 25) Nx high-entropy films by reactive sputtering. Entropy. **20**(9), 624 (2018)
31. Sajjad, S., et al.: Synergistic evaluation of AgO2 nanoparticles with ceftriaxone against CTXM and blaSHV genes positive ESBL producing clinical strains of Uro-pathogenic E. coli. IET Nanobiotechnol. **13**(4), 435–440 (2019)
32. Tan, Y.H., et al.: Application of RSM and Taguchi methods for optimizing the transesterification of waste cooking oil catalyzed by solid ostrich and chicken-eggshell derived CaO. Renew. Energy. **114**, 437–447 (2017)
33. Nan, Y., et al.: Production of biodiesel from microalgae oil (Chlorella protothecoides) by non-catalytic transesterification in supercritical methanol and ethanol: process optimization. J. Supercrit. Fluids. **97**, 174–182 (2015)
34. Shieh, C.-J., Liao, H.-F., Lee, C.-C.: Optimization of lipase-catalyzed biodiesel by response surface methodology. Bioresour. Technol. **88**(2), 103–106 (2003)
35. Tan, S.X., et al.: Two-step catalytic reactive extraction and transesterification process via ultrasonic irradiation for biodiesel production from solid Jatropha oil seeds. Chem. Eng. Process. Process. Intensific. **146**, 107687 (2019)
36. Das, S., et al.: Process optimization of biodiesel production using waste snail shell as a highly active nanocatalyst. Int. J. Energy Res. **2023**(1), 6676844 (2023)
37. Simsek, S., Uslu, S., Simsek, H.: Response surface methodology-based parameter optimization of single-cylinder diesel engine fueled with graphene oxide dosed sesame oil/diesel fuel blend. Energy AI. **10**, 100200 (2022)
38. Pali, H.S., et al.: Enhancement of combustion characteristics of waste cooking oil biodiesel using TiO2 nanofluid blends through RSM. Fuel. **331**, 125681 (2023)
39. Teoh, Y.H., et al.: Optimization of engine out responses with different biodiesel fuel blends for energy transition. Fuel. **318**, 123706 (2022)
40. Özgür, C.: Optimization of biodiesel yield and diesel engine performance from waste cooking oil by response surface method (RSM). Petrol. Sci. Technol. **39**(17-18), 683–703 (2021)
41. Singh, D., et al.: A review on feedstocks, production processes, and yield for different generations of biodiesel. Fuel. **262**, 116553 (2020)
42. Agarwal, A.K., et al.: Effect of fuel injection timing and pressure on combustion, emissions and performance characteristics of a single cylinder diesel engine. Fuel. **111**, 374–383 (2013)
43. Attia, A.M., Hassaneen, A.E.: Influence of diesel fuel blended with biodiesel produced from waste cooking oil on diesel engine performance. Fuel. **167**, 316–328 (2016)
44. Viswanathan, V.K., et al.: Hybrid optimization and modelling of CI engine performance and emission characteristics of novel hybrid biodiesel blends. Renew. Energy. **198**, 549–567 (2022)
45. Shrivastava, P., Verma, T.N., Pugazhendhi, A.: An experimental evaluation of engine performance and emission characteristics of CI engine operated with Roselle and Karanja biodiesel. Fuel. **254**, 115652 (2019)
46. Qi, D., et al.: Experimental studies on the combustion characteristics and performance of a direct injection engine fueled with biodiesel/diesel blends. Energ. Conver. Manage. **51**(12), 2985–2992 (2010)
47. Sayin, C., Gumus, M.: Impact of compression ratio and injection parameters on the performance and emissions of a DI diesel engine fueled with biodiesel-blended diesel fuel. Appl. Therm. Eng. **31**(16), 3182–3188 (2011)
48. Elumalai, P., et al.: Artificial neural networks model for predicting the behavior of different injection pressure characteristics powered by blend of biofuel-nano emulsion. Energy Sci. Eng. **10**(7), 2367–2396 (2022)
49. Wang, Q., et al.: Potential improvement in PM-NOX trade-off in a compression ignition engine by n-octanol addition and injection pressure. PRO. **9**(2), 310 (2021)
50. Ramachander, J., Gugulothu, S.K., Surya, M.S.: An experimental assessment on the influence of high fuel injection pressure with ternary fuel (diesel-Mahua methyl ester-Pentanol) on

performance, combustion and emission characteristics of common rail direct injection diesel engine. Heat Mass Transf. **57**, 2015–2027 (2021)
51. Bora, B.J., et al.: Effect of compression ratio on performance, combustion and emission characteristics of a dual fuel diesel engine run on raw biogas. Energ. Conver. Manage. **87**, 1000–1009 (2014)
52. Fahd, M.E.A., et al.: Experimental investigation of the performance and emission characteristics of direct injection diesel engine by water emulsion diesel under varying engine load condition. Appl. Energy. **102**, 1042–1049 (2013)
53. An, H., et al.: Combustion and emissions characteristics of diesel engine fueled by biodiesel at partial load conditions. Appl. Energy. **99**, 363–371 (2012)
54. Kannan, G., Anand, R.: Experimental investigation on diesel engine with diestrol–water micro emulsions. Energy. **36**(3), 1680–1687 (2011)
55. Ozcan, H.K.: Long term variations of the atmospheric air pollutants in Istanbul city. Int. J. Environ. Res. **9**(3), 781–790 (2012)
56. Simsek, S., et al.: Multi-objective-optimization of process parameters of diesel engine fueled with biodiesel/2-ethylhexyl nitrate by using Taguchi method. Energy. **231**, 120866 (2021)
57. Lertsathapornsuk, V., et al.: Microwave assisted in continuous biodiesel production from waste frying palm oil and its performance in a 100 kW diesel generator. Fuel Process. Technol. **89**(12), 1330–1336 (2008)
58. Gnanamoorthi, V., Devaradjane, G.: Effect of compression ratio on the performance, combustion and emission of DI diesel engine fueled with ethanol–Diesel blend. J. Energy Inst. **88**(1), 19–26 (2015)
59. Dharmegowda, I.Y., et al.: Optimization for waste coconut and fish oil derived biodiesel with MgO nanoparticle blend: Grey relational analysis, grey wolf optimization, driving training based optimization and election based optimization algorithm. Fuel. **338**, 127249 (2023)
60. Sivaramakrishnan, K., Ravikumar, P.: Optimization of operational parameters on performance and emissions of a diesel engine using biodiesel. Int. J. Environ. Sci. Technol. **11**, 949–958 (2014)
61. Devarajan, Y., et al.: Performance, combustion, and emission analysis of neat palm oil biodiesel and higher alcohol blends in a diesel engine. Energy Fuel. **31**(12), 13796–13801 (2017)
62. Yesilyurt, M.K., et al.: Investigation on the structural effects of the addition of alcohols having various chain lengths into the vegetable oil-biodiesel-diesel fuel blends: An attempt for improving the performance, combustion, and exhaust emission characteristics of a compression ignition engine. Fuel. **269**, 117455 (2020)
63. Singh, A., et al.: Optimization of performance and emission characteristics of CI engine fueled with Jatropha biodiesel produced using a heterogeneous catalyst (CaO). Fuel. **280**, 118611 (2020)
64. Man, X., et al.: Influence of engine load and speed on regulated and unregulated emissions of a diesel engine fueled with diesel fuel blended with waste cooking oil biodiesel. Fuel. **180**, 41–49 (2016)
65. Deepanraj, B., et al.: Comparison of jatropha and karanja biofuels on their combustion characteristics. Int. J. Green Energy. **14**(15), 1231–1237 (2017)
66. Kech, J., et al.: Exhaust Gas Recirculation: Internal Engine Technology for Reducing Nitrogen Oxide Emissions, Engine Technology, MTU Friedrichshafen GmbH, www.mtu-online.com (2011)
67. Bora, B.J., Saha, U.K.: Experimental evaluation of a rice bran biodiesel–biogas run dual fuel diesel engine at varying compression ratios. Renew. Energy. **87**, 782–790 (2016)

Chapter 4
Artificial Intelligence Modelling of Engine Performance and Emission Characteristics

4.1 Introduction

The biodiesel extracted from Garcinia Gummi-Gutta seeds was subjected to a transesterification process for biodiesel conversion. The biodiesel blends (biodiesel + diesel) were prepared and fuelled by a diesel engine. The combination of prepared biodiesel blends and engine variables was investigated experimentally using the design of experiments, and the corresponding performance and emission characteristics were recorded. Empirical equations were derived from the experimental input-output data sets. ANNs were used to develop the process models in forward and reverse directions. ANNs learn with huge data sets and the corresponding data have been collected by generating random data artificially. The data sets were used to train and optimize neural network parameters to reduce mean squared error to a minimal value. The models were trained by establishing forward and reverse models. The identical input-output data sets were used to develop ANN models to establish forward modelling (predicting output variables for known input variables) and reverse modelling (predicting input variables for desired output variables) tasks. The model predictions were validated through statistical tests against prediction accuracies with experimental cases. The prediction performances of developed models (BPNN and GA-NN) were compared to determine the best prediction model for practical applications. Figure 4.1 illustrates the framework for establishing the ANN model for predicting input-output variables.

M. Patel G. C. et al., *Biofuel Production, Performance, and Emission Optimization*, Springer Optimization and Its Applications 215,
https://doi.org/10.1007/978-3-031-84806-3_4

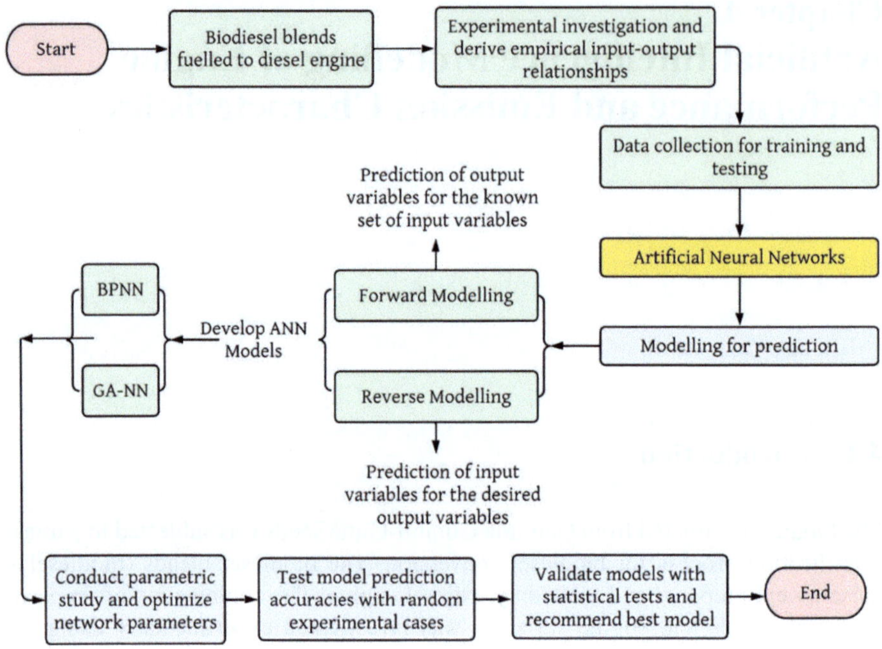

Fig. 4.1 The ANN model developed for predicting input-output variables

4.2 Advantages of Artificial Intelligence Over Statistical Design of Experiments

Artificial intelligence (AI) refers to a computer's capability to perform tasks using reasoning abilities, generalization, and prior experiences [1]. For AI to function autonomously, it must undergo training [2]. The system requires data input and instruction to analyse the issue, consider previous experiences and logical reasoning, and provide the most effective solution [3]. AI is extensively used on a global scale for many purposes, including health care [4], electoral forecasting [5], cybersecurity [6], traffic management [7], as well as the operation of self-governing robots both on land and underwater pipelines and cables [8]. Many methodologies are used to optimize the efficiency of AI. Artificial intelligence techniques are categorized as artificial neural networks (ANNs), machine learning (ML), fuzzy logic, and hybrid models [9]. AI techniques aid in minimizing the computing expenses associated with optimization [10, 11]. The system will receive many large data sets to execute complex operations and reduce errors and computing time. AI offers the advantage of automation [12]. ANNs are used to forecast the impact of engine parameters on the quality attributes of emissions and power generated [13, 14]. Biological-architectural connections influence it, and available data are utilized to educate a network with defined input parameters (neurons) and predict outputs [15]. ANNs may be used to construct a process model that captures linear and non-linear response correlations [16]. The network's capacity to store information encompasses the ability to handle incomplete input and perform parallel processing [17, 18]. An ANN is a very effective technique for addressing the limitations of low-order polynomial equations and dealing with noisy or incomplete data, resulting in improved predictive capabilities [19]. The objective of the ANN model is to forecast numerous outcomes concurrently. ANNs may be used for real-time and batch process monitoring [20]. The potential advantages of ANNs over DOE are listed as follows: handling complex large volumes of structured and unstructured data sets (including images, text, and sound) [21, 22], adaptive learning [23, 24], prediction [25–27], automation and efficiency [12, 28, 29], scalability to larger data sets [30, 31], real-time analysis [20], and so on.

4.3 Introduction to Artificial Neural Networks

The term "artificial intelligence" was coined in 1956 by John McCarthy [32]. A subset of artificial intelligence systems is artificial neural networks (ANNs). ANNs play various roles in biology, electronics, computer science, mathematics, and engineering [23]. The computational system simulates human brain function. Network training teaches the ANN how input and output patterns relate [33]. Developing and applying such a model is relatively simple, but it is more effective than other techniques. The system comprises an interconnected network of neurons that store information in its memory based on a predefined set of input and output patterns. If new knowledge is addressed to the exact location, the memory is updated without losing the old information [34]. The input and output layers constitute the two layers of a single-layer feed-forward network. The input layer receives input signals and transmits them to the output layer through weighted synaptic connections [35]. Computation occurs in the output layer, but information processing is unidirectional in a single-layer feed-forward neural network. A multilayer feed-forward neural network is a neural network architecture with multilayer layers. The input and output layers are separated by the hidden layer. Hidden layers acts as intermediary layers in the network, with each hidden layer containing a fixed number of neurons [36].

The connections between neurons in the input layer and the hidden layer are determined by input-hidden layer weights. Similarly, the weights associated with neurons in the output layer and the hidden layer are known referred to as hidden-output layer weights [37]. Hidden layers allow for hidden computations to be performed before transmitting the results from the input layer or other hidden layers to the output layer. Feedback loops are fundamental components of recurrent neural networks [38]. Output neurons provide feedback by sending their output to neurons in the previous or same layer [39, 40]. This feedback mechanism demonstrates the system's dynamic behaviour, characterized by non-linearity, and significantly influences the functional behaviour of the information processing system [41].

4.4 ANNs Modelling of Engine Performance and Emission Characteristics

Implementing stricter laws requires engines to function with enhanced cleanliness, while fierce market rivalry compels engines to be engineered for greater efficiency and decreased fuel consumption [42, 43]. Nevertheless, there is a trade-off between the amount of fuel used and the level of pollution control. Advancing the spark timing in spark ignition (SI) engines may reduce the levels of carbon monoxide (CO) and unburned hydrocarbon (HC) in the exhaust [44, 45]. This is achieved by providing more time for fuel to oxygenate. However, this may result in elevated nitrogen oxide (NOx) emissions [46].

On the other hand, delaying the ignition time may decrease the amount of nitrogen oxide (NOx) emissions. Still, it increases the levels of carbon monoxide (CO) and hydrocarbon (HC) emissions [47]. The calibration work conducted throughout the engine development process is comprehensive, necessitating adherence to both emission requirements and market demands.

Historically, dynamometer tests were used for engine mapping. To make this process more efficient, computational fluid dynamics (CFD) simulations, including both three-dimensional (3D) and one-dimensional (1D) models, have been used [48]. However, these models include constraints and require significant computing resources. Artificial intelligence is increasingly being seen as a favourable option because of its resilience compared to 1D simulations and its cost-effectiveness compared to 3D CFD models [49, 50]. Existing research demonstrates that machine learning models have shown to be very successful in accurately predicting a wide range of engine characteristics [51].

The objective is to evaluate the capability of ANNs, a machine learning technique, in predicting engine fuel consumption and emissions. This prediction is based on the engine speed and load parameters. The potential benefit of this strategy is minimizing the need for extensive experimental runs and computing expenses during engine mapping. ANNs have been extensively used in engine research among several machine learning methods [14, 52]. Prior research has effectively used ANNs to simulate the connections between factors such as fuel characteristics, thermodynamic properties, emissions, and NOx emissions. The research assesses the suitability of an ANN model implemented using the neural network toolbox in MATLAB 2020a [53]. Three primary contributions are emphasized: (i) The ANN-predicted engine reactions were compared to experimental results, showing a high level of agreement. The suggested approach offers expedited outcomes with comparable precision to multidimensional computational fluid dynamics (CFD) or comprehensive empirical investigations. (ii) Evaluating the model's accuracy on test samples, addressing the limitation of forecasting the map under complete working circumstances. The findings indicate that an ANN model can forecast a spark ignition engine's layout, enabling the enhancement of electric motors in hybrid automobiles. (iii) Addressing the need for more research on predicting fuel consumption and emissions under dynamic operating situations, hence filling a gap in the existing literature. The comparison between the dynamic performance predicted by ANN and the actual data demonstrates concurrence, serving as a benchmark for engine maps in hybrid cars. The following sections will comprehensively explain the data-collecting process, the architecture, and the hyper parameter configurations of the ANN model.

4.5 Data Collection for ANN Modelling

The input variables have significance in modelling the output variable(s) under consideration and are chosen using adequate decision-making methods. According to the literature review, throttle position, air flow mass, manifold pressure, power, engine temperature, air temperature, and exhaust gas temperature are used to predict the performance of the engine, such as brake power, torque, and engine emissions, such as CO, CO_2, HC, and NOx [54, 55].

The data set is divided into training and testing sets [56]. The neural network uses the training set to learn patterns, which serves as the primary data set. The testing set assesses a trained network's ability to generalize.

If the data used for model development falls within the range [X_{min}, X_{max}], test data must also fall within the same range [X_{min}, X_{max}]. Training data can also be generated within the same range. It is recommended to sample training data slightly beyond the data set's boundaries in the extreme regions. The goal is to improve the model's performance.

The input space usually has more than one variable. Each parameter may have a range used to gather information. In the same way, there could be more than one output, and each could have a different range. Because of this, each input and output needs its own set of factors that make them equal. The network will only work if the raw input and output data are processed first. A lot of work goes into making the prediction model.

There are two ways to prepare data for use. This is the normalization method, and this is the method for transformation. This method changes the raw data to be sent only to one network. But normalization changes a single input so that it is spread out evenly and in a range that the network can use. As we already said, one should know about the systems and processes the model should show. That way, one can better understand how the process works. This will help the network learn better, making it work better.

As the last step in pre-processing data, check the data to ensure it is correct. The goal is to move the data around so that it is spread out evenly across the range of data. This is true for both data that comes in and data that leaves. If data is scaled, the range of the input and output data should match the range of the neuron, which is what adds and squeezes the values. Normalization is a necessary last step before any other change.

One can scale anything in the range of the raw data. However, most of the literature uses a min-max function, the below function, to smooth the data to be between 0.1 and 0.9 [57].

$$X_n = \frac{(X - \mathrm{X}) * 0.8}{(X_{\max} - X_{\min})} + 0.1$$

The defined value is shown above as X_n, the present data value; ($X_{\min}$), the least significant number in the range; and ($X_{\max}$), the most important number.

After training to a certain skill level, the network should give the user simple-to-understand output values. To do this, one has to process the already collected data. The following equation is used to get the network's outputs from the normalized form.

$$X = \frac{(X_{\max} - X_{\min}) * (X_n - 0.1)}{0.8} + X_{\min}$$

4.5.1 *Training Data*

Training and testing data are fundamental to developing a reliable and effective neural network model. Training data is used to teach the model the underlying patterns and relationships within the data, allowing it to learn from examples. This is analogous to the educational phase, where the model is exposed to various scenarios and learns to predict outcomes based on input variables. The quality and diversity of the training data directly influence the model's ability to generalize to new, unseen data. To this end, experimental data (comprising 27 sets) and their derived empirical regression equations were used to generate a supplementary set of 973 artificially randomized data points for the training data set.

4.5.2 *Testing Data*

Testing data serves as a benchmark to evaluate the model's predictive performance. This step gauges how well the model has learned and its ability to predict outcomes. After the learning phase, it is crucial that the model can accurately generalize to the responses specific to the testing data set, not just the training data. These data sets ensure that a neural network model is knowledgeable, adaptable, and robust enough for deployment in real-world applications. Therefore, the developed models were tested using 20 random cases not included in the training phase.

4.6 Method of ANN Modelling

The multilayer perceptron (MLP) has proven instrumental in resolving contentious and complex technological issues. Using a successful error backpropagation mechanism during training allows for the sequential resolution of these issues. Learning rules like Gradient Decent, Adaptive Filtering, or the Least Mean Square algorithm fix mistakes during learning. The design has three different kinds of layers: input, concealed, and output. The term "Feed-Forward Neural Network" comes from signals in the future from the input layers to the output layers. The network will have no feedback links; therefore, the processed data will go through many layers. Figure 4.2 schematically depicts a multilayer perceptron feed-forward neural network with input and output parameters. The configured neural network must produce the expected output from the given set of inputs. There are two methods to adjust the weights. Using prior information to set the weights is one approach. Another option is to use the data to train a model to learn the output pattern using predefined rules. There are two modes that the backpropagation algorithm may run in for a particular training epoch: sequential mode and batch mode. The pattern basis determines the neurons' weights when operating in sequential mode. Batch mode, on the other hand, relies only on an epoch basis for adjusting synaptic weights and bias. When training a backpropagation model, the sequential mode is often used. It is essential to train networks so there are little deviations from the objective and expected values. The number of epochs and the magnitude of the error may be

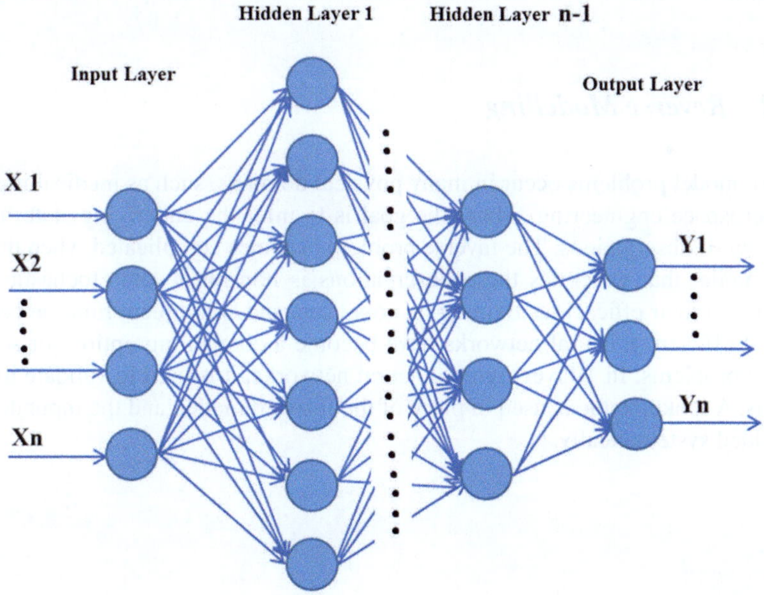

Fig. 4.2 Feed-forward architecture of ANNs

specified for error specifications. With epoch specification, testing the data begins after the training data has run for the given number of epochs. When an error criterion is applied, the training data will be processed until the stated value of the error is reached.

4.6.1 Forward Modelling

The forward model is a fundamental form of a neural network that processes input in a unidirectional manner. This kind of artificial neural network called a feed-forward neural network, has nodes without feedback loops, so data flows only forward and never backward [58]. This architecture, also known as a multilayer neural network, allows information to move from input to output nodes via hidden layers. One notable feature that sets this network configuration apart is the lack of links that permit data to be returned from the output node.

The following is how a feed-forward neural network approximates functions:

- Classifiers are computed using an algorithmic formula $y = f \times (x)$.
- As a result, input x is classified under category y.
- The feed-forward model defines y as $f(x; \theta)$. This number indicates the closest approximation to the function.

The network gains knowledge of the connection between input and output pairs. ANNs are used to tackle a variety of issues, including filtering, optimization, pattern recognition, completion association, parameter estimation, and prediction [59].

4.6.2 Reverse Modelling

Inverse model problems occur in many physical domains, such as medical imaging and aerospace engineering, where the goal is to infer the underlying state from a given set of observations. The inverse problem becomes complicated when the forward model that generates these observations is non-linear and stochastic [60]. Because of their efficacious training on noisy data and consequent runtime computational efficiency, neural networks have become a compelling option for solving inverse problems. In 1-layer sigmoid neural networks, it is used to validate inverse models. A lookup table is used in place of the forward model, and the input domain is gridded systematically.

4.7 Back Propagation Neural Network (BPNN)

At present, backpropagation is the most commonly used neural network architecture. This is because backpropagation networks can quickly learn complex multidimensional mappings. The back propagation neural network architecture's guiding principles cover learning approaches, function approximation capability, performance assessment, and architectural design [53]. Figure 4.3 shows the flow diagram of the working mechanism of backpropagation.

Feed-forward computations frequently produce an output value that differs from the intended or target teacher output value, which leads to an error. An error value is generated by this discrepancy between the target and actual feed-forward values. The neural network model aims to maximize its predictive output by reducing this error rate. This optimization process is made possible through the use of backpropagation.

4.7.1 Weights

The learning process of a neural network is heavily impacted by the initialization of its weights, which are usually set to small, random values. The learning process might need correction if the starting weights are equal, but the best solution is uneven weights. As a result of either stabilizing errors or experiencing an increase in errors during training, the network may be unable to effectively learn from the training data if this deviation is present. Negative weight drifting can occur if the network is trained beyond a low plateau region, leading to an increase in error and a decrease in the network's mapping quality. It is recommended to start training with a new set of random weights to fix this issue [61].

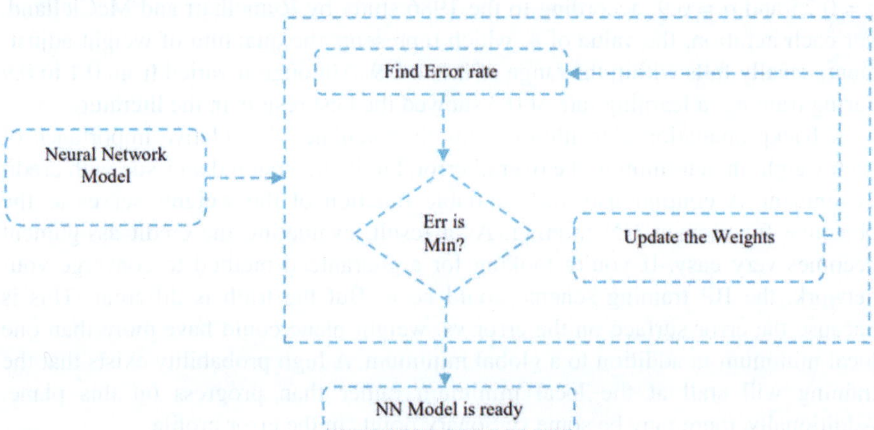

Fig. 4.3 Flow diagram of the working mechanism of backpropagation

4.7.2 Hidden Layers and Neurons

The network's capacity to generalize is derived from the hidden layer. Theoretically, a single hidden layer with sufficient hidden neurons can approximate a continuous function [61, 62]. In actuality, neural networks with just one hidden layer are far more common than those with two. The researchers considered a network with two hidden layers. Optimal guidelines for choosing the amount of hidden neurons were not provided. The calculation of the hidden neurons was done using thumb rules. The number of hidden neurons with n inputs and m output neurons for the three-layer neural network can be found by taking the smallest square root of $n*m$ neurons. Hence, the minimum hidden layer starts with four neurons and keeps increasing the number of neurons until the optimized value (The time taken to learn must be minimum). If the number of neurons is more, then knowing the input-output relationship will take longer.

4.7.3 Learning Rate and Momentum Constant

The amount of weight adjustment done at each step is also controlled by the learning rate parameter (η), which controls the network's convergence rate. If this parameter is not chosen correctly, convergence problems may occur. The learning rate coefficient determines the efficiency of a backpropagation algorithm. No data is available to help choose the learning rate parameter depending on the relationship type in the data sets [63]. Understanding the learning rate coefficient thus becomes an issue-specific iterative process. Setting the learning rate parameter too low will cause incredibly slow convergence. Excessively high values of η could cause the network to oscillate or diverge, leading to non-convergence. In this setting, learning will not be possible. The majority of calculations yield satisfactory results when $\eta = 0.25$ and $\alpha = 0.9$, according to the 1986 study by Rumelhart and McClelland. For each iteration, the value of η, which represents the quantum of weight adjustment, usually falls within the range of 0.1 to 0.9. Although it varied from 0.1 to 0.9 during training, a learning rate of 0.3 showed the best results in the literature.

A backpropagation algorithm essentially calculates the relative importance of each weight in determining the overall error, but it encountered an issue with credit assignment. A continuously differentiable function of the weights serves as the objective function in NN training. As a result, evaluating the credit assignment becomes very easy. If you're looking for a guaranteed method to converge your network, the BP training scheme could be it. But the truth is different. This is because the error surface on the error vs. weight plane could have more than one local minimum in addition to a global minimum. A high probability exists that the training will stall at the local minimum rather than progress on this plane. Additionally, there may be some stationary points in the error profile.

Additionally, the stationary points prevent the BP algorithm from learning. An FFNN's training scheme incorporates the momentum term to evade these issues and speed up the network's convergence. The technique entails adding a portion of the weight adjustment from the prior time step to the current weight adjustments. As per the present weight adjustment, it is structured as follows.

$$\left(w(t) = -\mu\Delta E(t) + \alpha\Delta w(t-1)\right)$$

The variables t and $(t-1)$ in the given equation represent the most recent and current training steps, respectively. The momentum factor, also known as the momentum term is another name for the term α that you might hear. By including the momentum term, the descent is accelerated even more, which is a positive outcome. By applying this process, the network can be helped to escape local minima and move down the slope of the error profile. From the previous equation, we can deduce that the magnitude of the momentum term increases the impact on convergence. In most cases, α falls between 0.1 and 0.9. During training, researchers tested momentum factor values between 0.1 and 0.9 and found that 0.5 yields the best results.

4.7.4 Constants of Activation Function

The output of a processing node is defined by mathematical formulas called activation functions. The net input of the unit is activated by applying the activation function. Non-linear functions, such as tanh or log sigmoid, can be used as activation functions [64]. Overly large values can cause the network to paralyse or show signs of inhibited training. Therefore, transfer functions are used to avoid these kinds of problems. Network learning often uses the sigmoidal transfer function because of its non-linearity and continuous differentiability. The most popular function is the sigmoid function, a continuous unipolar function. The sigmoid function is $f(x)$ is defined $1/(1 + e$^$(-x))$.

One can choose an activation function for the outputs that fits the distribution of the target values. The logistic function performs very well when dealing with binary numbers (0/1). Consequently, a logistic (sigmoid) function-based activation function has been utilized.

4.7.5 Bias and Variance

The simplest definition of bias is the model's incapacity to account for discrepancies or errors between its anticipated and real values. Error or bias refers to the variations between actual or expected values and the forecasted ones [65].

Most AI systems and algorithms rely on data for training, as they are data-driven. Therefore, data is essential for the functioning of these algorithms and systems. The algorithms trained on the underlying training data will detect and incorporate any biases present in that data into their predictions. Consequently, pre-existing biases in the data can affect algorithms that use it, leading to biased results. Biases in the data may already exist and can be amplified by algorithms [66]. Additionally, even when the data is unbiased, algorithms may exhibit biased behaviour due to certain design decisions. The outputs of these biased algorithms outputs can be integrated into real systems and influence users choices, creating more biased data that will be used to train algorithms in the future.

A systematic inaccuracy known as bias arises when the machine learning process makes incorrect assumptions.

$$\text{Bias}(y) = f(y) - y$$

Here, "$f(y)$" represents the estimator y's predicted value. The model's fit to the data is determined by measurement. A low bias value indicates fewer assumptions were made when building the target function, whereas a high bias value indicates that more assumptions were made.

Variance in machine learning refers to how training on various subsets of the training data alters a predictive model's performance [67]. More specifically, variance describes the variability of the model with respect to its sensitivity to a distinct subset of the training data set, or, more accurately, its degree of adaptability to the new subset.

Let y represent the target variable's actual values and $\hat{y}$ represent its projected values. A model's variance can then be computed using the expected value of the square of the difference between the predicted values and the expected value of the predicted values [68].

$$\text{Variance} = \left[\left(\hat{y} - \mathrm{f}(\hat{y})\right)^2\right]$$

Where $f(\hat{y})$ is the value you expect from the predicted numbers. The predicted value is found by taking the mean of all the training data. When the variance is low, the model is less affected by changes in the training data. On the other hand, it is very affected by changes in the training data when the variance is high.

The bias-variance trade-off is influenced by the level of "complexity exhibited by the function class f [69]. The concept is that by increasing the complexity of the model, the bias will reduce as it allows for better estimates of *f*opt(); however, it will also raise the variance as the magnitude of oscillations around the optimal model will grow. Therefore, the function class f is selected to minimize the total of the squared bias and its variance, effectively optimizing the balance between bias and variance.

4.8 Genetic Algorithm–Neural Network (GA-NN)

In 1992, J.H. Holland proposed Genetic Algorithm (GA) which mimics biological evolution and the Darwinian Theory of survival of the fittest. GA operators such as crossover, and recombination, mutation, and selection are used to solve complex problems, leading to optimized and efficient solutions [70].

ANNs have several characteristics similar to the human brain. The input, hidden, and output layers refer to the various layers that process items assembled into a network that runs parallel to each other. Data is sent to the network from the input layer, transferred through hidden layers, first layer to the hidden layer, and produces output at the output layer [71].

Hybrid GA-NN has been utilized for prediction and classification applications, using an ANN as the main modelling tool and a GA to optimize the ANN topology or connected weights [72] The quantity of neurons in the network and their arrangement within the layers significantly impact the effectiveness of the entire [73]. Two issues are associated with hybrid systems [72]:

(i) The majority of hybrid systems prioritize categorization over effective feature selection. Before classification, large amounts of data are filtered, imputed, and normalized to remove undesirable data and improve performance.
(ii) There is no clear explanation for the architecture of hybrid systems. ANN and GA configurations, such as the number of nodes, activation function, GA population size, chromosomal count, and GA evaluation operator, are unknown, leading to erroneous findings.

4.8.1 Genetic Algorithm Operators: Selection, Mutation, and Crossover

Genetic algorithms have three main operations:

(a) Selection
(b) Crossover
(c) Mutation

4.8.1.1 Selection

It is based on the survival of the fittest concept. It is an important step in deciding whether a string will participate in reproduction. The string with the maximum fitness value is considered for the next step in the algorithm [74].

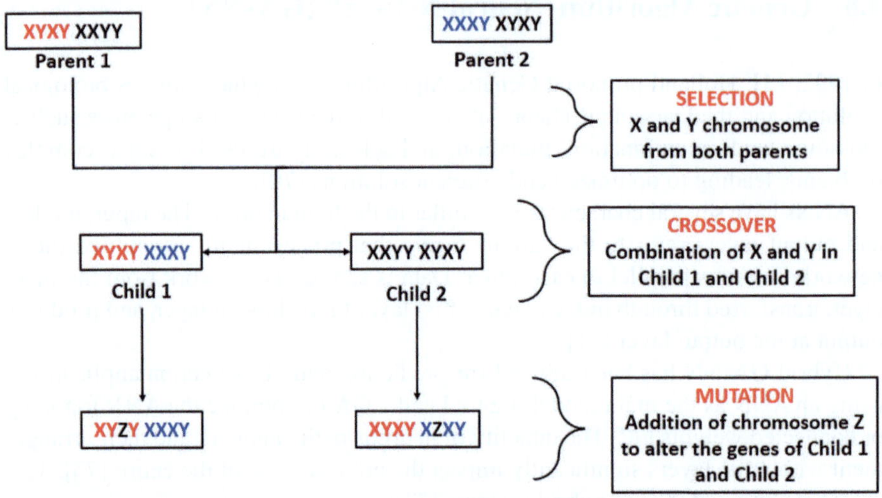

Fig. 4.4 Basic explanation of GA operators: Selection, Crossover, and Mutation

4.8.1.2 Crossover

Crossovers connect two or more parent chromosomes together to form new, efficient ones. After selecting pairs of parent chromosomes, crossover enables parents to exchange information to generate children [75]. These approaches are parent-centric and mean-centric operators [76]. In the parent-centric approach, a pair of strings mate together to form the base for the generation of new strings while in the mean-centric approach, a common place is chosen for the pair of strings for a crossover [77].

4.8.1.3 Mutation

The mutation operation changes progeny genes and diversifies the population [76], leading to a solution outside the space. Figure 4.4 shows the basic concept of selection, crossover, and mutation.

4.9 Model Evaluation

Models were developed to predict random test cases in both forward and reverse directions. The predicted models were evaluated with metrics that quantify the accuracies. Two commonly employed metrics used are:

1. Percent deviation (PD) in prediction: PD measures the differences between predicted and experimental values expressed as percentage of the experimental value. The computation is done using the following Eq. (4.1). The PD value, i.e. 0 indicates a perfect prediction, while positive or negative values indicate overestimation or underestimation, respectively.

$$\mathrm{PD} = \frac{(\text{Model predicted value} - \text{Experimental value})}{\text{Experimental value}} \times 100\% \tag{4.1}$$

2. Mean Absolute Percentage Error (MAPE): The average of absolute percentage errors of prediction measures the MAPE. The computation is done using the formulae given in Eq. (4.2).

$$\mathrm{MAPE} = \left(\frac{1}{n}\right)\sum_{i=1}^{n}\left|\frac{\text{Experimental value of } i\text{th observation} - \text{Predicted value of } i\text{th observation}}{\text{Experimental value of } i\text{th observation}}\right| \times 100\% \tag{4.2}$$

Lower values of MAPE indicate better predictive accuracy, with 0 being the ideal case where the predictions perfectly match the actual or experimental values.

4.10 Results of Modelling Tools and Techniques

ANNs were applied to develop models and establish relationships in forward and reverse directions. In forward directions, the neurons of input (such as engine load, injection pressure, compression ratios, and blend type) layers were aimed to predict the outputs (i.e. BSFC, CO, UHC, BTE, and NOx). In reverse directions, ANNs were developed to predict the inputs for the desired outputs. The prediction performances of ANNs (BPNN and GA-NN) were compared in reverse modelling, whereas ANNs (BPNN and GA-NN) were compared with CCD models to carry out the forward modelling tasks.

4.10.1 Results of Forward Modelling

The predictions made by ANNs were dependent on the quality of training and testing data. Therefore, 1000 sets of data (27 experimental and 973 artificially randomized data) were utilized for training and 20 sets of randomly generated values of experimental variables were used for model evaluation (see Table 4.1). The results of the BPNN and GA-NNs models developed for predictions are discussed in Table 4.1.

Table 4.1 Input-output data for test cases

	Input data				Output data				
Exp. no	EL (%)	IP (bar)	CR	BT (vol. %)	BTE (%)	BSFC (kg/ kw.h)	CO (% vol.)	UHC (ppm)	NOx (ppm)
1	43	254	15	25	22.47	0.41	0.165	29.14	746.3
2	52	259	14	16	25.24	0.343	0.253	45.16	383
3	74	183	16	21	27.56	0.384	0.246	38.93	674.3
4	54	227	17	26	24.22	0.347	0.181	29.18	873.4
5	22	216	16	26	21.12	0.396	0.14	21.86	793.9
6	49	178	17	24	24.64	0.432	0.189	30.95	765.1
7	72	202	14	18	27.71	0.401	0.289	47.09	517.7
8	48	208	15	11	25.16	0.307	0.348	55.93	280.7
9	76	235	15	28	25.42	0.38	0.219	35.32	944.9
10	40	223	14	19	23.22	0.36	0.2	37.23	456.2
11	28	255	14	17	24.1	0.449	0.186	37.13	400.5
12	66	180	14	23	25.35	0.425	0.238	39.8	681.8
13	63	254	17	10	27	0.28	0.311	53.31	302.1
14	27	242	14	16	23.99	0.424	0.199	39.11	383.8
15	68	254	16	21	27.06	0.383	0.217	35.87	670.1
16	95	218	15	21	30.82	0.359	0.253	40.99	694.8
17	62	199	15	24	24.29	0.352	0.214	35.39	731.6
18	28	233	17	10	24.98	0.32	0.289	45.91	292.7
19	26	241	16	25	22.48	0.437	0.142	23.41	790.4
20	82	219	16	19	29.07	0.332	0.245	40.67	613.2

4.10.1.1 BPNN

The 1000 training input-output data sets were passed to the neural network in batch mode. Training performance is evaluated based on the mean squared error (MSE). Training aims to reduce MSE to a minimum value.

The MSE value is reliant on neural network parameters such as hidden neurons, learning rate for the hidden layer (η_1), learning rate for the output layer (η_2), momentum constant (α), activation function constant for the hidden layer (a_H), output layer-1 (a_{O1}), output layer-2 (a_{O2}) and bias (b) value. To date no universal and acceptable standards have been defined to optimize training parameters. Therefore, authors adopted varying one parameter at a time after keeping all other parameters at their respective middle values between their operating levels. For example, varying hidden neurons between 4 and 25 led to determining the optimal neurons for the hidden layers after keeping the fixed middle value of η_1, η_2, α, a_H, a_{O1}, a_{O2}, and b value equal to 0.0445, 0.0445, 0.0445, 5.5, 5.5, and 0.0000505, respectively. The MSE variations with network parameters were presented in Fig. 4.5. Once all the network parameters were optimized the corresponding MSE values of bias parameters at the end of training were treated as the optimal choice. The systematically

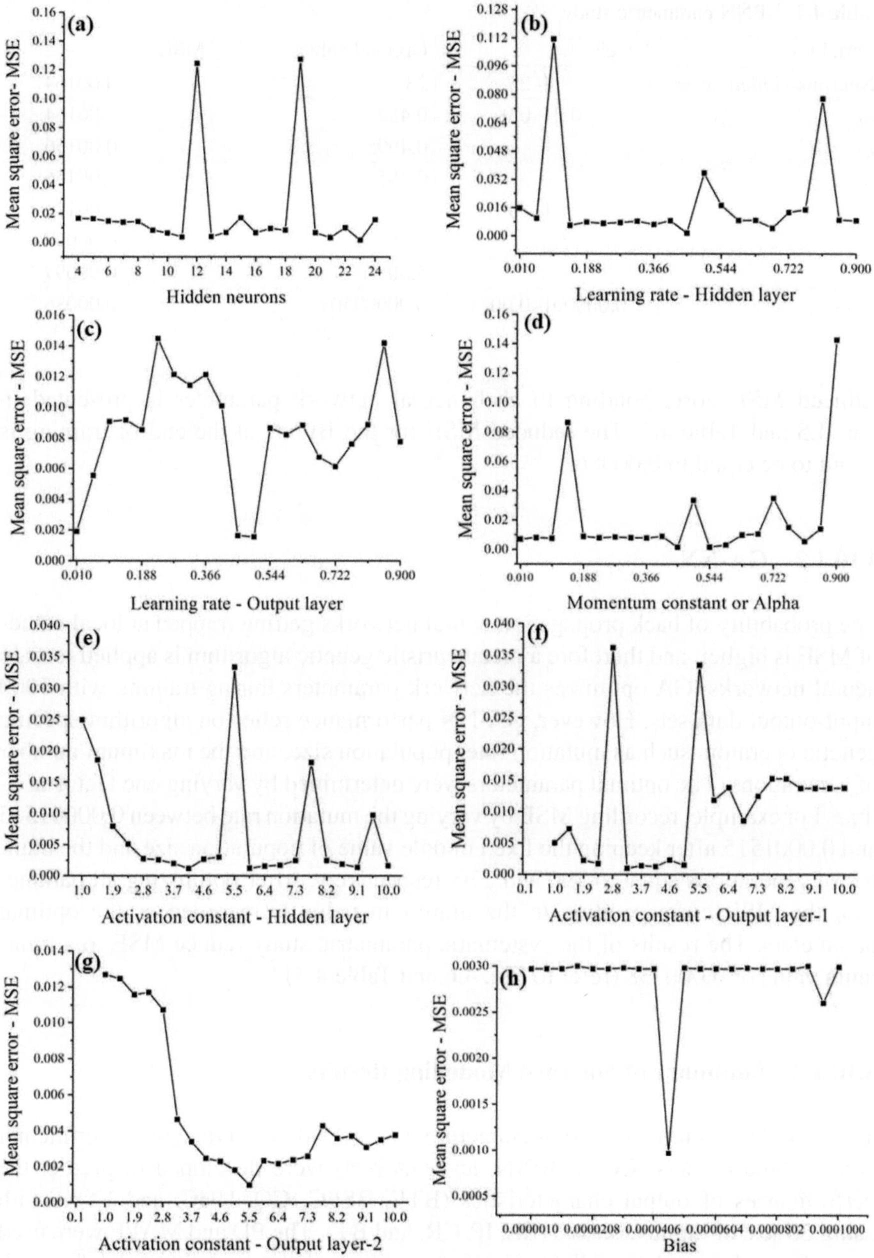

Fig. 4.5 Parametric study of BPNN: (**a–h**) MSE vs. hidden neurons, learning rate (hidden and output layer), alpha, activation constant (hidden and output layers), and bias value

Table 4.2 BPNN parametric study

Variables	Levels	Optimal values	MSE
Neurons-hidden layer	4–25	23	0.00164
η_1	0.01–0.9	0.455	0.00164
η_2		0.4995	0.00156
A		0.455	0.00156
a_H	1–10	4.15	0.00121
a_{O1}		3.25	0.00097
a_{O2}		5.50	0.00097
Bias	0.000001–0.0001	0.00009505	0.00086

reduced MSE corresponding to each neural network parameter is presented in Fig. 4.5 and Table 4.2. The reduced MSE for the BPNN at the end of training is found to be equal to 0.00086.

4.10.1.2 GA-NN

The probability of back propagation neural networks getting trapped at local values of MSE is higher, and therefore a metaheuristic genetic algorithm is applied to train neural networks. GA optimizes the network parameters during training with 1000 input-output data sets. However, GA-NN performance relies on algorithm-specific genetic operators such as mutation rate, population size, and the maximum number of generations. The optimal parameters were determined by varying one factor-at-a-time. For example, recording MSE by varying the mutation rate between 0.00001515 and 0.0001515 after keeping the fixed middle value of population size and the number of generations equal to 190 and 270, respectively. After optimizing all parameters, the MSE corresponding to the minimum value is recorded as the optimal parameters. The results of the systematic parametric study reduce MSE to a minimum value of 0.000738 (refer to Fig. 4.6 and Table 4.3).

4.10.1.3 Summary of Forward Modelling Results

The optimized neural network architecture were tested on 20 random experimental cases. Three models (CCD, BPNN, and GA-NN) were developed to predict the performances of output characteristics (BTE, BSFC, CO, UHC, and NOx) with using 20 sets of input variables (EL, IP, CR, and BT). The PD and MAPE were used to evaluate the model prediction performances, and the best fit model for each response was selected. The PD values for predicting the 20 random experimental cases by CCD, BPNN, and GA-NN ranged from −1.66% to 3.01%, −1.29% to 2.22%, and −1.38% to 1.85% for BTE, respectively. For BSFC, the ranges were for

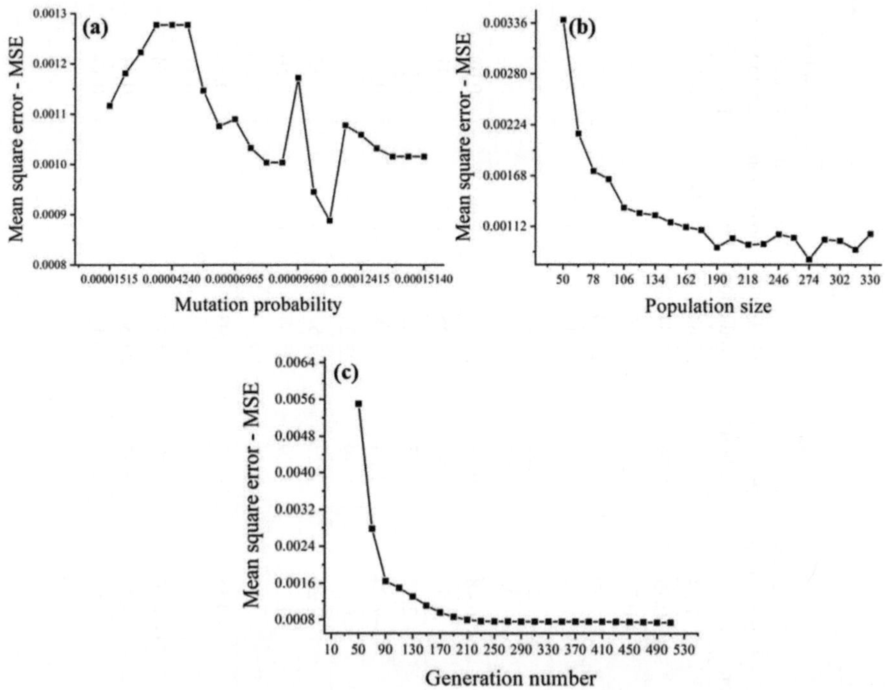

Fig. 4.6 Parametric study of GA-NN: (**a–c**) MSE vs. mutation probability, population size, and generation number

Table 4.3 GA-NN parametric study

Variables	Levels	Optimal values	MSE
Mutation	0.00001515–0.0001515	0.0001106	0.000888
Population size	50–330	274	0.000758
Generations	10–530	490	0.000738

CO, −4.88% to 6.50%, −3.66% to 7.00%, and −3.16% to 6.50% for UHC, and −1.29% to 2.74%, −1.16% to 2.24 and −0.9% to 1.6% for NOx, respectively (see Tables 4.4, 4.5, 4.6, 4.7, and 4.8; Fig. 4.7).

It is noteworthy that all three models followed similar trends for the 20 data points on both positive and negative sides of the reference zero line. The BPNN and GA-NN models predicted closer to the reference zero line for all responses compared to the statistical CCD models (see Fig. 4.7).

The MAPE values for predicting all 20 random experimental cases by CCD, BPNN, and GA-NN were 1.38%, 1.07%, and 0.69% for BTE, 6.92%, 5.40%, and 4.06% for BSFC, 3.76%, 3.15%, and 2.63% for CO, 0.97%, 0.79%, and 0.66% for UHC, and 2.96%, 2.48%, and 2.24% for NOx, respectively (see Tables 4.4, 4.5, 4.6, 4.7, and 4.8; Fig. 4.8). The grand average absolute PD for predicting all responses for 20 data points by the three models was 3.198% for CCD, 2.578% for BPNN, and

Table 4.4 Models' prediction performances for response: BTE

Models	CCD				BPNN			GA-NN		
Test case	Experimental value	CCD prediction	Deviation in prediction (%)	Absolute percent deviation in prediction	Model prediction	Deviation in prediction (%)	Absolute percent deviation in prediction	Model prediction	Deviation in prediction (%)	Absolute percent deviation in prediction
1	22.47	22.87	1.78	1.780	22.77	1.34	1.34	22.63	0.71	0.71
2	25.24	25.54	1.19	1.189	25.44	0.79	0.79	25.34	0.40	0.40
3	27.56	27.16	−1.45	1.451	27.356	−0.74	0.74	27.45	−0.40	0.40
4	24.22	24.62	1.65	1.652	24.68	1.90	1.90	24.53	1.28	1.28
5	21.12	21.72	2.84	2.841	21.52	1.89	1.89	21.43	1.47	1.47
6	24.64	24.24	−1.62	1.623	24.34	−1.22	1.22	24.3	−1.38	1.38
7	27.71	27.31	−1.44	1.444	27.41	−1.08	1.08	27.51	−0.72	0.72
8	25.16	25.36	0.79	0.795	25.28	0.48	0.48	25.18	0.08	0.08
9	25.42	25.52	0.39	0.393	25.55	0.51	0.51	25.45	0.12	0.12
10	23.22	23.92	3.01	3.015	23.72	2.15	2.15	23.43	0.90	0.90
11	24.1	23.7	−1.66	1.660	23.8	−1.24	1.24	23.88	−0.91	0.91
12	25.35	25.15	−0.79	0.789	25.25	−0.39	0.39	25.2	−0.59	0.59
13	27	27.4	1.48	1.481	27.6	2.22	2.22	27.5	1.85	1.85
14	23.99	23.79	−0.83	0.834	23.68	−1.29	1.29	23.87	−0.50	0.50
15	27.06	26.76	−1.11	1.109	26.96	−0.37	0.37	27.13	0.26	0.26
16	30.82	30.52	−0.97	0.973	30.65	−0.55	0.55	30.93	0.36	0.36
17	24.29	24.69	1.65	1.647	24.49	0.82	0.82	24.39	0.41	0.41
18	24.98	24.58	−1.60	1.601	24.68	−1.20	1.20	24.77	−0.84	0.84
19	22.48	22.28	−0.89	0.890	22.58	0.44	0.44	22.38	−0.44	0.44
20	29.07	28.97	−0.34	0.344	28.87	−0.69	0.69	29.12	0.17	0.17
Minimum value of deviation (%)			−1.66			−1.29			−1.38	
Maximum value of deviation (%)			3.01			2.22			1.85	
Average absolute deviation in prediction (%)				1.38	1.07			0.69		

Table 4.5 Models' prediction performances for BSFC

Model	CCD				BPNN			GA-NN		
Test case	Experimental value	CCD prediction	Deviation in prediction (%)	Absolute percent deviation in prediction	Model prediction	Deviation in prediction (%)	Absolute percent deviation in prediction	Model prediction	Deviation in prediction (%)	Absolute percent deviation in prediction
1	0.41	0.414	0.98	0.976	0.416	1.46	1.46	0.415	1.22	1.22
2	0.343	0.373	8.75	8.746	0.365	6.41	6.41	0.358	4.37	4.37
3	0.384	0.354	−7.81	7.813	0.366	−4.69	4.69	0.371	−3.39	3.39
4	0.347	0.387	11.53	11.527	0.377	8.65	8.65	0.371	6.92	6.92
5	0.396	0.426	7.58	7.576	0.388	−2.02	2.02	0.408	3.03	3.03
6	0.432	0.392	−9.26	9.259	0.406	−6.02	6.02	0.41	−5.09	5.09
7	0.401	0.361	−9.98	9.975	0.368	−8.23	8.23	0.381	−4.99	4.99
8	0.307	0.327	6.51	6.515	0.322	4.89	4.89	0.316	2.93	2.93
9	0.38	0.39	2.63	2.632	0.394	3.68	3.68	0.388	2.11	2.11
10	0.36	0.406	12.78	12.778	0.398	10.56	10.56	0.387	7.50	7.50
11	0.449	0.409	−8.91	8.909	0.414	−7.80	7.80	0.418	−6.90	6.90
12	0.425	0.405	−4.71	4.706	0.402	−5.41	5.41	0.409	−3.76	3.76
13	0.28	0.268	−4.29	4.286	0.264	−5.71	5.71	0.272	−2.86	2.86
14	0.424	0.404	−4.72	4.717	0.408	−3.77	3.77	0.412	−2.83	2.83
15	0.383	0.353	−7.83	7.833	0.359	−6.27	6.27	0.363	−5.22	5.22
16	0.359	0.329	−8.36	8.357	0.337	−6.13	6.13	0.341	−5.01	5.01
17	0.352	0.392	11.36	11.364	0.382	8.52	8.52	0.375	6.53	6.53
18	0.32	0.311	−2.81	2.813	0.324	1.25	1.25	0.328	2.50	2.50
19	0.437	0.417	−4.58	4.577	0.419	−4.12	4.12	0.425	−2.75	2.75
20	0.332	0.322	−3.01	3.012	0.324	−2.41	2.41	0.328	−1.20	1.20
Minimum value of deviation (%)			−9.98			−8.23			−6.90	
Maximum value of deviation (%)			12.78			10.56			7.50	
Average absolute deviation in prediction (%)				6.92	5.40			4.06		

Table 4.6 Models' prediction performances for CO

Model	CCD				BPNN			GA-NN		
Test case	Experimental value	CCD prediction	Deviation in prediction (%)	Absolute percent deviation in prediction	Model prediction	Deviation in prediction (%)	Absolute percent deviation in prediction	Model prediction	Deviation in prediction (%)	Absolute percent deviation in prediction
1	0.165	0.173	4.85	4.848	0.171	3.64	3.64	0.17	3.03	3.03
2	0.253	0.242	−4.35	4.348	0.244	−3.56	3.56	0.245	−3.16	3.16
3	0.246	0.234	−4.88	4.878	0.237	−3.66	3.66	0.239	−2.85	2.85
4	0.181	0.191	5.52	5.525	0.19	4.97	4.97	0.189	4.42	4.42
5	0.14	0.149	6.43	6.429	0.147	5.00	5.00	0.146	4.29	4.29
6	0.189	0.182	−3.70	3.704	0.183	−3.17	3.17	0.184	−2.65	2.65
7	0.289	0.28	−3.11	3.114	0.287	−0.69	0.69	0.286	−1.04	1.04
8	0.348	0.356	2.30	2.299	0.358	2.87	2.87	0.357	2.59	2.59
9	0.219	0.226	3.20	3.196	0.225	2.74	2.74	0.224	2.28	2.28
10	0.2	0.213	6.50	6.500	0.214	7.00	7.00	0.213	6.50	6.50
11	0.186	0.18	−3.23	3.226	0.181	−2.69	2.69	0.182	−2.15	2.15
12	0.238	0.23	−3.36	3.361	0.232	−2.52	2.52	0.233	−2.10	2.10
13	0.311	0.321	3.22	3.215	0.319	2.57	2.57	0.318	2.25	2.25
14	0.199	0.203	2.01	2.010	0.202	1.51	1.51	0.201	1.01	1.01
15	0.217	0.21	−3.23	3.226	0.211	−2.76	2.76	0.213	−1.84	1.84
16	0.253	0.246	−2.77	2.767	0.248	−1.98	1.98	0.249	−1.58	1.58
17	0.214	0.224	4.67	4.673	0.222	3.74	3.74	0.22	2.80	2.80
18	0.289	0.28	−3.11	3.114	0.281	−2.77	2.77	0.283	−2.08	2.08
19	0.142	0.146	2.82	2.817	0.147	3.52	3.52	0.146	2.82	2.82
20	0.245	0.25	2.04	2.041	0.249	1.63	1.63	0.248	1.22	1.22
Minimum value of deviation (%)			−4.88			−3.66			−3.16	
Maximum value of deviation (%)			6.50			7.00			6.50	
Average absolute deviation in prediction (%)				3.76	3.15			2.63		

Table 4.7 Models' prediction performances for UHC

Model	CCD				BPNN			GA-NN		
Test Case	Experimental value	CCD prediction	Deviation in prediction (%)	Absolute percent deviation in prediction	Model prediction	Deviation in prediction (%)	Absolute percent deviation in prediction	Model prediction	Deviation in prediction (%)	Absolute percent deviation in prediction
1	29.14	29.54	1.37	1.373	29.47	1.13	1.13	29.41	0.93	0.93
2	45.16	45.46	0.66	0.664	45.4	0.53	0.53	45.34	0.40	0.40
3	38.93	38.53	−1.03	1.027	38.59	−0.87	0.87	38.67	−0.67	0.67
4	29.18	29.58	1.37	1.371	29.51	1.13	1.13	29.45	0.93	0.93
5	21.86	22.46	2.74	2.745	22.35	2.24	2.24	22.21	1.60	1.60
6	30.95	30.55	−1.29	1.292	30.59	−1.16	1.16	30.67	−0.90	0.90
7	47.09	46.69	−0.85	0.849	46.8	−0.62	0.62	46.89	−0.42	0.42
8	55.93	56.13	0.36	0.358	56.19	0.46	0.46	56.27	0.61	0.61
9	35.32	35.42	0.28	0.283	35.48	0.45	0.45	35.52	0.57	0.57
10	37.23	37.93	1.88	1.880	37.87	1.72	1.72	37.81	1.56	1.56
11	37.13	36.73	−1.08	1.077	36.79	−0.92	0.92	36.83	−0.81	0.81
12	39.8	39.6	−0.50	0.503	39.62	−0.45	0.45	39.64	−0.40	0.40
13	53.31	53.71	0.75	0.750	53.39	0.15	0.15	53.45	0.26	0.26
14	39.11	38.91	−0.51	0.511	38.99	−0.31	0.31	39.01	−0.26	0.26
15	35.87	35.57	−0.84	0.836	35.67	−0.56	0.56	35.71	−0.45	0.45
16	40.99	40.69	−0.73	0.732	40.71	−0.68	0.68	40.73	−0.63	0.63
17	35.39	35.79	1.13	1.130	35.71	0.90	0.90	35.65	0.73	0.73
18	45.91	45.51	−0.87	0.871	45.59	−0.70	0.70	45.64	−0.59	0.59
19	23.41	23.21	−0.85	0.854	23.27	−0.60	0.60	23.29	−0.51	0.51
20	40.67	40.57	−0.25	0.246	40.59	−0.20	0.20	40.64	−0.07	0.07
Minimum value of deviation (%)			−1.29			−1.16			−0.90	
Maximum value of deviation (%)			2.74			2.24			1.60	
Average absolute deviation in prediction (%)				0.97	0.79			0.66		

Table 4.8 Models' prediction performances for NOx

Model	CCD				BPNN			GA-NN		
Test case	Experimental value	CCD prediction	Deviation in prediction (%)	Absolute percent deviation in prediction	Model prediction	Deviation in prediction (%)	Absolute percent deviation in prediction	Model prediction	Deviation in prediction (%)	Absolute percent deviation in prediction
1	746.3	766.3	2.68	2.680	762.8	2.21	2.21	761.5	2.04	2.04
2	383	393	2.61	2.611	389.5	1.70	1.70	388.2	1.36	1.36
3	674.3	644.3	−4.45	4.449	647.8	−3.93	3.93	652.2	−3.28	3.28
4	873.4	886.4	1.49	1.488	882.9	1.09	1.09	881.6	0.94	0.94
5	793.9	811.9	2.27	2.267	808.4	1.83	1.83	806.7	1.61	1.61
6	765.1	745.1	−2.61	2.614	748.9	−2.12	2.12	752.5	−1.65	1.65
7	517.7	505.7	−2.32	2.318	509.2	−1.64	1.64	510.5	−1.39	1.39
8	280.7	288.7	2.85	2.850	285.2	1.60	1.60	283.9	1.14	1.14
9	944.9	972.9	2.96	2.963	969.4	2.59	2.59	968.1	2.46	2.46
10	456.2	473.2	3.73	3.726	469.7	2.96	2.96	468.4	2.67	2.67
11	400.5	382.5	−4.49	4.494	384.5	−4.00	4.00	387.3	−3.30	3.30
12	681.8	671.8	−1.47	1.467	675.3	−0.95	0.95	676	−0.85	0.85
13	302.1	324.1	7.28	7.282	320.6	6.12	6.12	319.3	5.69	5.69
14	383.8	351.8	−8.34	8.338	356.3	−7.17	7.17	356.6	−7.09	7.09
15	670.1	658.1	−1.79	1.791	660.6	−1.42	1.42	662.9	−1.07	1.07
16	694.8	686.8	−1.15	1.151	683.3	−1.66	1.66	682	−1.84	1.84
17	731.6	738.6	0.96	0.957	735.1	0.48	0.48	733.8	0.30	0.30
18	292.7	286.7	−2.05	2.050	283.2	−3.25	3.25	281.9	−3.69	3.69
19	790.4	780.4	−1.27	1.265	783.9	−0.82	0.82	785.2	−0.66	0.66
20	613.2	598.2	−2.45	2.446	600.7	−2.04	2.04	602	−1.83	1.83
Minimum value of deviation (%)			−8.34			−7.17			−7.09	
Maximum value of deviation (%)			7.28			6.12			5.69	
Average absolute deviation in prediction (%)				2.96	2.48			2.24		

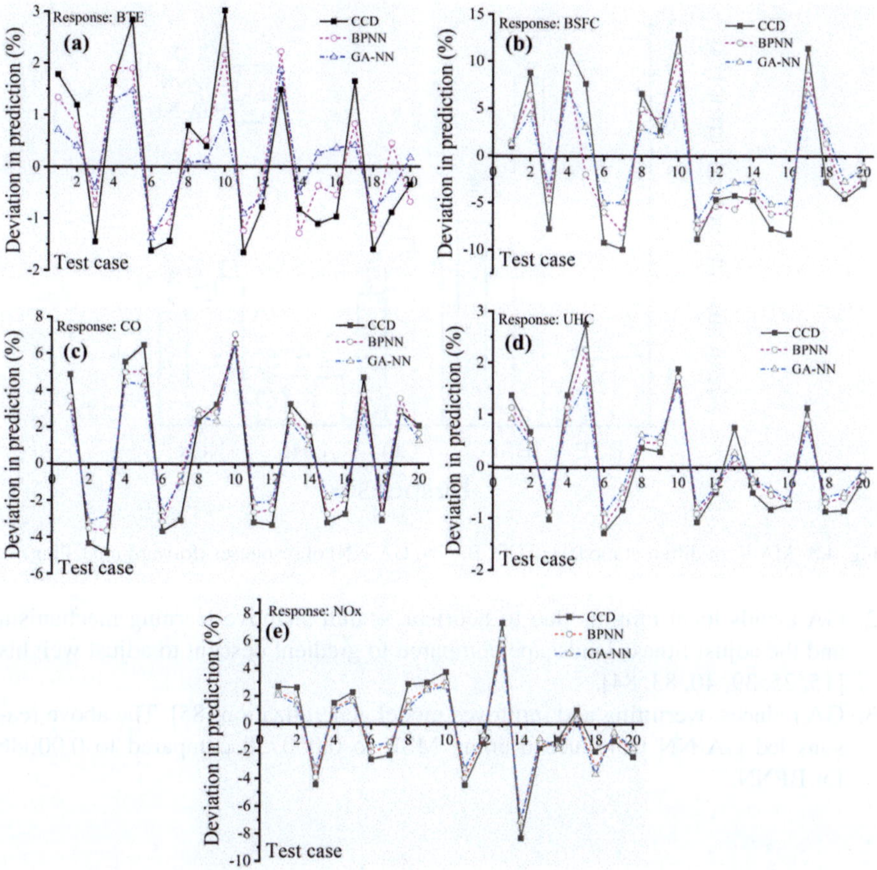

Fig. 4.7 Forward modelling prediction performances of three models (CCD, BPNN, GA-NN) for different responses: (**a**) BTE, (**b**) BSFC, (**c**) CO, (**d**) UHC, (**e**) NOx

2.056% for GA-NN. The better prediction by neural models over CCD can be attributed to several reasons

1. Higher degree of adaptability [78].
2. Ability to capture and model complex non-linear relationships [79].
3. Learn in an iterative approach [80].
4. Neural network learn from large data sets and therefore have a greater ability to capture the process dynamics with complex data sets that are missed with smaller data sets modelled by CCD [81].

Furthermore, GA-NN predicted better than BPNN in most of the data points, probably due to the following:

1. GA systematically searches for global solutions in large space at multiple spatial locations and fine tune hyper parameters (such as η and α) [82].

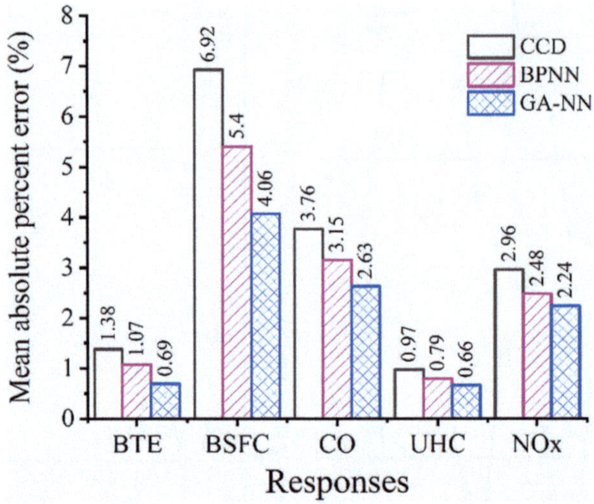

Fig. 4.8 MAPE of different models (CCD, BPNN, GA-NN) of responses (forward modelling)

2. GA avoids local minima due to heuristic search adaptive learning mechanism and the adjust fitness landscape compared to gradient descent to adjust weights [15, 25, 39, 40, 83, 84].
3. GA reduces overfitting and improves model generalization [85]. The above reasons led GA-NN to reduce training MSE to 0.000738 compared to 0.00086 for BPNN.

4.10.2 Results of Reverse Modelling

Reverse modelling was conducted to determine the required sets of input parameters (EL, IP, CR, and BT) for the desired outputs (BTE, BSFC, CO, NOx, and UHC). Neural network models (BPNN and GA-NN) were applied for the reverse modelling task. The CCD model is not used for reverse predictions primarily due to the following reasons:

1. CCD is a parametric model designed to develop explicit quadratic relationships that are inherently irreversible.
2. The CCD model cannot deal with redundancy and over determination in data sets comprising multiple inputs and outputs (whereas the CCD model develops direct and identical relationships between inputs and a single output), which is an essential requirement for reverse modelling.
3. CCD relies on predetermined relationships and is not a learning-based model, whereas neural models learn directly by training on reversed input-output pairs.

The limitations of the CCD model can be effectively tackled with data-driven learning models like neural networks. The neural models developed for predictions were tested with 20 random experimental cases. Note that the input-output data sets used for training and testing for forward modelling remain the same for reverse modelling. The neural models (BPNN and GA-NN) were compared, and the best model for reverse modelling was determined.

4.10.2.1 BPNN

The BPNN model is applied to predict the multiple inputs (EL, CR, IP, and BT) simultaneously with a set of desired outputs (BTE, BSFC, CO, NOx, and UHC). The neural network architecture designed for forward modelling remains identical for reverse modelling, except that training and testing of the models were carried out in reversed input-output pairs. The training and network parameter study were conducted similarly to those used during forward modelling. The optimized BPNN parameters and resulting MSE are presented in Table 4.9. The BPNN parameters reduce the MSE to a minimal value equal to 0.00487.

4.10.2.2 GA-NN

In GA-NN, GA replaces the BP algorithm to train the optimized structure of BPNN. Before prediction, GA-NN parameters were optimized with the 1000 training data sets supplied in batch mode. The optimized GA-NN parameters (mutation probability, population size, and generations) reduce the MSE to a minimal value equal to 0.00212. GA-NN parameter study results are presented in Table 4.10.

4.10.2.3 Summary Results of Reverse Modelling

In reverse modelling, 20 random experimental cases were used to evaluate the model prediction performances. The performances of the BPNN and GA-NN models were compared. The predicted responses (EL, CR, IP, and BT) by BPNN and GA-NN both models followed the similar trend line, with data points distributed on both positive and negative sides of the reference zero line (Figs. 4.9, 4.10, 4.11, and 4.12).

Table 4.9 BPNN parametric study-reverse modelling

Variables	Levels	Optimal values	MSE
Neurons-hidden layer	4–25	21	0.01250
η_1	0.01–0.9	0.366	0.01070
η_2		0.455	0.00906
A		0.4995	0.00753
a_H	1–10	2.80	0.00637
a_{O1}		1.90	0.00595
a_{O2}		6.40	0.00531
Bias	0.000001–0.0001	0.0000703	0.00487

Table 4.10 GA-NN parametric study-reverse modelling

Variables	Levels	Optimal values	MSE
Mutation	0.00001515–0.0001515	0.0001174	0.000888
Population size	50–330	260	0.00212
Generations	10–530	270	0.00212

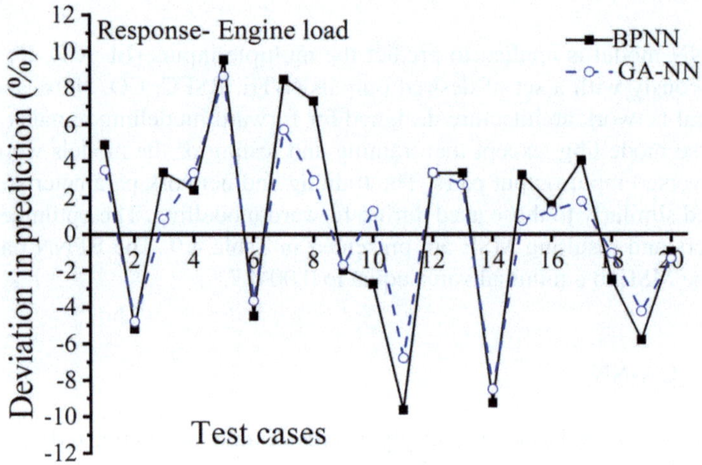

Fig. 4.9 Engine load prediction tests of different models

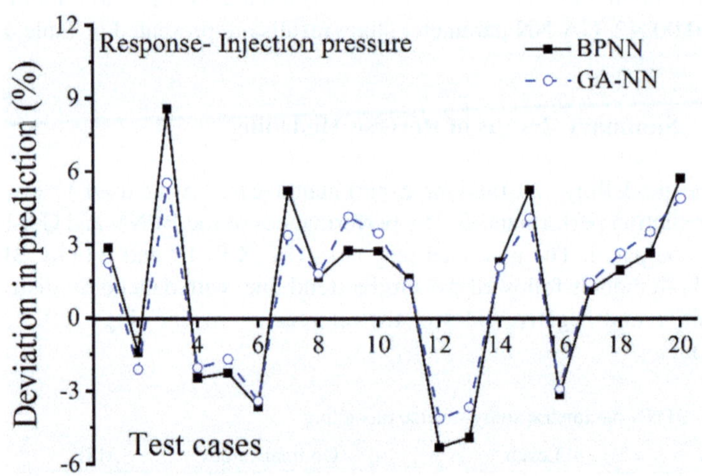

Fig. 4.10 Injection pressure prediction tests of different models

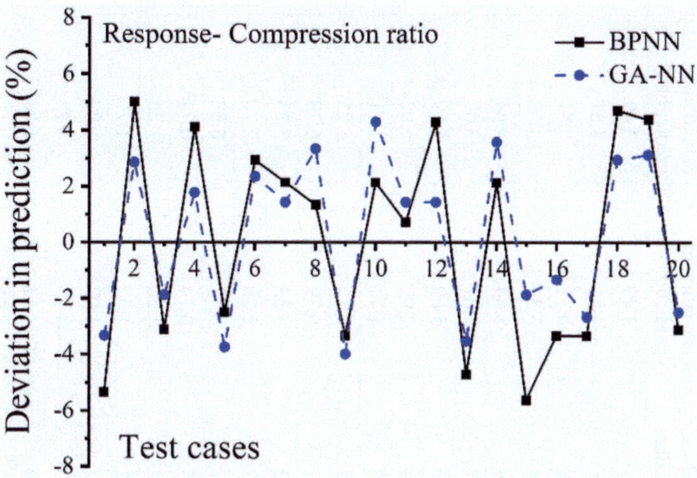

Fig. 4.11 Compression ratio prediction tests of different models

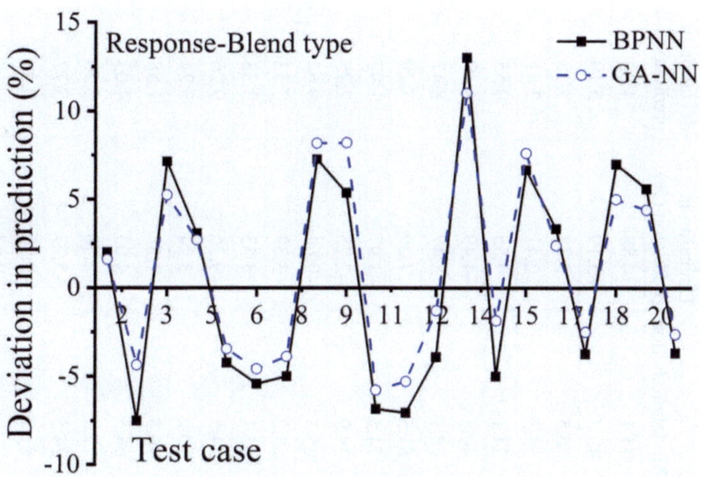

Fig. 4.12 Blend type prediction tests of different models

The percentage differences (PD) in predicting the 20 random experimental cases by BPNN and GA-NN models varied in the ranges of [−9.64% to 9.55%, −8.52% to 8.64%] for EL, [−5.33% to 8.58%, −4.17% to 5.52%] for IP, [−5.63% to 5.0%, −4.0% to 4.29%] for CR, and [−7.50% to 13.0%, −5.79% to 11.0%] for BT (refer to Tables 4.11, 4.12, 4.13, and 4.14; Figs. 4.9, 4.10, 4.11, and 4.12. The prediction ranges of the GA-NN model were closer to the reference zero line than those of BPNN, indicating better performance. This difference could be attributed to the error surface at the end of training, which was lower for GA-NN (0.00212)

Table 4.11 Models' prediction performances for engine load

Model	BPNN				GA-NN		
Test case	Experimental value	Model prediction	Deviation in prediction (%)	Absolute percent deviation in prediction	Model prediction	Deviation in prediction (%)	Absolute percent deviation in prediction
1	43	45.1	4.88	4.88	44.5	3.49	3.49
2	52	49.3	−5.19	5.19	49.5	−4.81	4.81
3	74	76.5	3.38	3.38	74.6	0.81	0.81
4	54	55.3	2.41	2.41	55.8	3.33	3.33
5	22	24.1	9.55	9.55	23.9	8.64	8.64
6	49	46.8	−4.49	4.49	47.2	−3.67	3.67
7	72	78.1	8.47	8.47	76.1	5.69	5.69
8	48	51.5	7.29	7.29	49.4	2.92	2.92
9	76	74.5	−1.97	1.97	74.7	−1.71	1.71
10	40	38.9	−2.75	2.75	40.5	1.25	1.25
11	28	25.3	−9.64	9.64	26.1	−6.79	6.79
12	66	68.2	3.33	3.33	68.2	3.33	3.33
13	63	65.1	3.33	3.33	64.7	2.70	2.70
14	27	24.5	−9.26	9.26	24.7	−8.52	8.52
15	68	70.2	3.24	3.24	68.5	0.74	0.74
16	95	96.5	1.58	1.58	96.2	1.26	1.26
17	62	64.5	4.03	4.03	63.1	1.77	1.77
18	28	27.3	−2.50	2.50	27.7	−1.07	1.07
19	26	24.5	−5.77	5.77	24.9	−4.23	4.23
20	82	81.2	−0.98	0.98	80.9	−1.34	1.34
Minimum value of deviation (%)			−9.64			−8.52	
Maximum value of deviation (%)			9.55			8.64	
Average absolute deviation in prediction (%)				4.70	3.40		

Table 4.12 Models' prediction performances for injection pressure

Model	BPNN				GA-NN		
Test case	Experimental value	Model prediction	Deviation in prediction (%)	Absolute percent deviation in prediction	Model prediction	Deviation in prediction (%)	Absolute percent deviation in prediction
1	254	261.3	2.87	2.87	259.7	2.24	2.24
2	259	255.3	−1.43	1.43	253.5	−2.12	2.12
3	183	198.7	8.58	8.58	193.1	5.52	5.52
4	227	221.4	−2.47	2.47	222.3	−2.07	2.07
5	216	211.1	−2.27	2.27	212.3	−1.71	1.71
6	178	171.5	−3.65	3.65	171.9	−3.43	3.43
7	202	212.5	5.20	5.20	208.8	3.37	3.37
8	208	211.4	1.63	1.63	211.7	1.78	1.78
9	235	241.5	2.77	2.77	244.7	4.13	4.13
10	223	229.1	2.74	2.74	230.7	3.45	3.45
11	255	259.1	1.61	1.61	258.6	1.41	1.41
12	180	170.4	−5.33	5.33	172.5	−4.17	4.17
13	254	241.5	−4.92	4.92	244.6	−3.70	3.70
14	242	247.5	2.27	2.27	247	2.07	2.07
15	254	267.3	5.24	5.24	264.3	4.06	4.06
16	218	211.1	−3.17	3.17	211.6	−2.94	2.94
17	199	201.2	1.11	1.11	201.8	1.41	1.41
18	233	237.5	1.93	1.93	239.1	2.62	2.62
19	241	247.4	2.66	2.66	249.5	3.53	3.53
20	219	231.5	5.71	5.71	229.7	4.89	4.89
Minimum value of deviation (%)			−5.33			−4.17	
Maximum value of deviation (%)			8.58			5.52	
Average absolute deviation in prediction (%)				3.38	3.03		

Table 4.13 Models' prediction performances for compression ratio

Model	BPNN				GA-NN		
Test case	Experimental value	Model prediction	Deviation in prediction (%)	Absolute percent deviation in prediction	Model prediction	Deviation in prediction (%)	Absolute percent deviation in prediction
1	15	14.2	−5.33	5.33	14.5	−3.33	3.33
2	14	14.7	5.00	5.00	14.4	2.86	2.86
3	16	15.5	−3.13	3.13	15.7	−1.88	1.88
4	17	17.7	4.12	4.12	17.3	1.76	1.76
5	16	15.6	−2.50	2.50	15.4	−3.75	3.75
6	17	17.5	2.94	2.94	17.4	2.35	2.35
7	14	14.3	2.14	2.14	14.2	1.43	1.43
8	15	15.2	1.33	1.33	15.5	3.33	3.33
9	15	14.5	−3.33	3.33	14.4	−4.00	4.00
10	14	14.3	2.14	2.14	14.6	4.29	4.29
11	14	14.1	0.71	0.71	14.2	1.43	1.43
12	14	14.6	4.29	4.29	14.2	1.43	1.43
13	17	16.2	−4.71	4.71	16.4	−3.53	3.53
14	14	14.3	2.14	2.14	14.5	3.57	3.57
15	16	15.1	−5.63	5.63	15.7	−1.88	1.88
16	15	14.5	−3.33	3.33	14.8	−1.33	1.33
17	15	14.5	−3.33	3.33	14.6	−2.67	2.67
18	17	17.8	4.71	4.71	17.5	2.94	2.94
19	16	16.7	4.38	4.38	16.5	3.13	3.13
20	16	15.5	−3.13	3.13	15.6	−2.50	2.50
Minimum value of deviation (%)			−5.63			−4.00	
Maximum value of deviation (%)			5.00			4.29	
Average absolute deviation in prediction (%)				3.42	2.67		

Table 4.14 Models' prediction performances for blend type

Model	BPNN				GA-NN		
Test case	Experimental value	Model prediction	Deviation in prediction (%)	Absolute percent deviation in prediction	Model prediction	Deviation in prediction (%)	Absolute percent deviation in prediction
1	25	25.5	2.00	2.00	25.4	1.60	1.60
2	16	14.8	–7.50	7.50	15.3	–4.38	4.38
3	21	22.5	7.14	7.14	22.1	5.24	5.24
4	26	26.8	3.08	3.08	26.7	2.69	2.69
5	26	24.9	–4.23	4.23	25.1	–3.46	3.46
6	24	22.7	–5.42	5.42	22.9	–4.58	4.58
7	18	17.1	–5.00	5.00	17.3	–3.89	3.89
8	11	11.8	7.27	7.27	11.9	8.18	8.18
9	28	29.5	5.36	5.36	30.3	8.21	8.21
10	19	17.7	–6.84	6.84	17.9	–5.79	5.79
11	17	15.8	–7.06	7.06	16.1	–5.29	5.29
12	23	22.1	–3.91	3.91	22.7	–1.30	1.30
13	10	11.3	13.00	13.00	11.1	11.00	11.00
14	16	15.2	–5.00	5.00	15.7	–1.88	1.88
15	21	22.4	6.67	6.67	22.6	7.62	7.62
16	21	21.7	3.33	3.33	21.5	2.38	2.38
17	24	23.1	–3.75	3.75	23.4	–2.50	2.50
18	10	10.7	7.00	7.00	10.5	5.00	5.00
19	25	26.4	5.60	5.60	26.1	4.40	4.40
20	19	18.3	–3.68	3.68	18.5	–2.63	2.63
Minimum value of deviation (%)			–7.50			–5.79	
Maximum value of deviation (%)			13.00			11.00	
Average absolute deviation in prediction (%)				5.64	4.60		

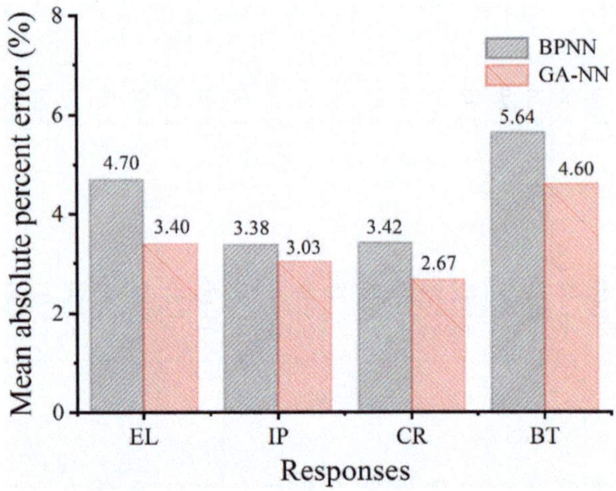

Fig. 4.13 MAPE of different models for responses (reverse modelling)

compared to BPNN (0.00487). This suggests that BPNN may get trapped at local minima, while GA-NN's heuristic learning mechanism helps avoid this issue. For the 20 random experimental cases, the PD for EL, IP, CR, and BT were 4.7% and 3.4%, 3.38%, and 3.03%, 3.42% and 2.67%, and 5.64% and 4.6% for BPNN and GA-NN, respectively (refer to Tables 4.11, 4.12, 4.13, and 4.14; Fig. 4.13). The grand average absolute PD for all outputs was 3.83% for BPNN and 3.03% for GA-NN. The prediction performances of reverse modelling were found to be inferior to those of forward modelling, possibly due to the complex interplay of input variables resulting in the same output.

From an industrial perspective, predicting variables like EL, CR, IP, and BT for desired performance metrics is crucial. DOE models and analyses a single output simultaneously, which might fail to capture the interdependencies of outputs and predict multiple outputs. The predictions are made with empirical regression equations using DOE, which require the transformation matrix to be square for reverse prediction. The transformation matrix derived from non-linear models might be non-invertible due to multi-collinearity (when input variables are highly correlated) among variables. Multi-collinearity leads to a singular matrix that prevents the reverse model from providing a single or any solutions. Empirical equations with only linear terms can be helpful for reverse prediction; however, including interaction terms leads to imprecise input-output relationships and reduces prediction precision. On the contrary, ANNs are trained with massive input-output data sets, allowing them to map complex input-output relationships. Furthermore, ANNs can predict inputs from outputs without inverting a transformation matrix. These capabilities make artificial neural networks suitable for reverse prediction.

The results of reverse modelling can be valuable for industrial applications.

4.11 Concluding Remarks

The predictive computational tools were developed for conducting biodiesel research in modelling both forward and reverse directions. A forward modelling task is carried out to predict the outputs (BTE, BSFC, CO, NOx, and UHC) for the known set of inputs (EL, CR, IP, and BT). A reverse modelling task is carried out to predict the inputs (EL, CR, IP, and BT) required for the desired output (BTE, BSFC, CO, NOx, and UHC), the primary requirement for industry personnel for online monitoring and process control. Three models were developed for forward modelling, and two neural models were used for reverse modelling tasks. The following conclusions were drawn from the present work:

1. A CCD-based experimental matrix was planned for four input parameters (EL, CR, IP, and BT), allowing output data collection (BTE, BSFC, CO, NOx, and UHC). The derived empirical equations from the experimental data collected were found to have a better coefficient of determination with a value greater than 0.95. Neural network models require massive data sets for training, and those were generated artificially with random values of an operating range of variables. The 1000 data sets used for neural network training comprised 27 experimental cases and 973 randomly generated data.
2. BPNN and GA-NN models trained with 1000 data sets supplied in batch mode. The BPNN parameters (hidden neurons, η_1, η_2, α, a_H, a_{O1}, a_{O2}, and b value) and GA-NN parameters (mutation probability, population size, and generations) were optimized to reduce the training error with the 1000 input-output data sets. GA-NN replaces the backpropagation algorithm that optimizes the BPNN parameters of neural networks. The network parameter study remains identical for both forward and reverse modelling, whereas the data pairs for input and output layers are interchanged. Forward modelling aims to predict the outputs (BTE, BSFC, CO, NOx, and UHC) from the known sets of inputs (EL, CR, IP, and BT). Reverse modelling predicts the inputs (EL, CR, IP, and BT) for the desired output (BTE, BSFC, CO, NOx, and UHC).
3. In forward modelling, tuning BPNN and GA-NN parameters reduced the MSE to a minimal value equal to 0.00086 and 0.000738. The PD in predicting 20 random experimental cases by CCD, BPNN and GA-NN were found to vary in the ranges of [−1.66 to 3.01%, −1.29 to 2.22%, −1.38 to 1.85%] for BTE, [−9.98 to 12.78%, −8.23 to 10.56%, −6.90 to 7.50%] for BSFC, [−4.88 to 6.50%, −3.66 to 7.00%, −3.16 to 6.50%] for CO, [−1.29 to 2.74%, −1.16 to 2.24%, −0.9 to 1.6%] for UHC, and [−8.34 to 7.28%, −7.17 to 6.12%, −7.09 to 5.69%] for NOx, respectively. The grand average absolute PD in predicting all responses for 20 data points by the three models was found to be equal to 3.198% for CCD, 2.578% for BPNN, and 2.056% for GA-NN, respectively. GA-NN model predicted better than BPNN and CCD considering all responses.
4. In reverse modelling, the BPNN and GA-NN parameters reduced MSE to a minimal value of 0.00487 and 0.00212, respectively. The PD in predicting 20 random experimental cases by BPNN and GA-NN models were found to vary in the

ranges of [−9.64 to 9.55%, −8.52 to 8.64%] for EL, [−5.33 to 8.58%, −4.17 to 5.52%] for IP, [−5.63 to 5.0%, −4.0 to 4.29%] for CR, [−7.50 to 13.0%, −5.79 to 11.0%] for BT, respectively. The grand average absolute PD in predicting all outputs (EL, IP, CR, and BT) by two models was equal to 3.83% for BPNN and 3.03% for GA-NN, respectively. GA-NN prediction performances were better than those obtained for BPNN in reverse modelling.

5. The prediction capabilities of reverse modelling were found to be lower than those of forward modelling because many sets of combinations of input variables produced identical results (due to the interplay of input variables). Stated differently, a production is produced when a parameter is increased together with a drop in another parameter and vice versa.
6. The developed forward and reverse models with various anticipated outputs and inputs guarantee adaptive training of the models, which may facilitate online monitoring with less reliance on industry workers.

Scope for Future Work

To date, no universally acceptable method has been established in the literature regarding the supremacy of algorithm performances across all areas or problems to be solved. The prediction performances of the neural networks rely on algorithmic performances in optimizing network architecture parameters during training. Therefore, the authors are working with other metaheuristic algorithms (such as TLBO, JAYA, PSO, DTBO, EBOA, and GWO) to tune neural network parameters towards the possibility of determining global solutions.

References

1. Shabbir, J., Anwer T.: Artificial intelligence and its role in near future. arXiv preprint arXiv:.01396 (2018)
2. Bitterman, D.S., Aerts, H.J., Mak, R.H.: Approaching autonomy in medical artificial intelligence. Lancet Dig. Health. **2**(9), e447–e449 (2020)
3. Nilsson, N.J.: Principles of Artificial Intelligence. Springer (1982)
4. Jiang, F., et al.: Artificial intelligence in healthcare: past, present and future. Stroke Vasc. Neurol. **2**(4) (2017)
5. dos Santos Brito, K., Adeodato, P.J.L.: Predicting Brazilian and US elections with machine learning and social media data. In 2020 International Joint Conference on Neural Networks (IJCNN). IEEE (2020)
6. Taddeo, M.: Three ethical challenges of applications of artificial intelligence in cybersecurity. Minds Mach. **29**, 187–191 (2019)
7. Gangwani, D., Gangwani, P.: Applications of machine learning and artificial intelligence in intelligent transportation system: a review. Applications of Artificial Intelligence Machine Learning: Select Proceedings of ICAAAIML, pp. 203–216 (2021)
8. Sharma, N., Pandey, J.K., Mondal, S.: A review of mobile robots: applications and future prospect. Int. J. Prec. Eng. Manufact. **24**(9), 1695–1706 (2023)
9. Bagheri, M., Akbari, A., Mirbagheri, S.A.: Advanced control of membrane fouling in filtration systems using artificial intelligence and machine learning techniques: a critical review. Process Saf. Environ. Protect. **123**, 229–252 (2019)

10. Islam, J., et al.: A holistic review on artificial intelligence techniques for well placement optimization problem. Adv. Eng. Softw. **141**, 102767 (2020)
11. Salehi, S., Schmeink A.: Data-centric green artificial intelligence: a survey. IEEE Transactions on Artificial Intelligence (2023)
12. Jha, N., Prashar, D., Nagpal, A.: Combining artificial intelligence with robotic process automation—an intelligent automation approach. In: Deep Learning Big Data for Intelligent Transportation: Enabling Technologies Future Trends, pp. 245–264 (2021)
13. Yusri, I., et al.: A review on the application of response surface method and artificial neural network in engine performance and exhaust emissions characteristics in alternative fuel. Renew. Sustain. Energy Rev. **90**, 665–686 (2018)
14. Hoang, A.T., et al.: A review on application of artificial neural network (ANN) for performance and emission characteristics of diesel engine fueled with biodiesel-based fuels. Sustain. Energy Technol. Assessm. **47**, 101416 (2021)
15. Prabhakar, D., et al.: A review of optimization and measurement techniques of the friction stir welding (FSW) process. J. Manufact. Mater. Process. **7**(5), 181 (2023)
16. Almeida, J.S.: Predictive non-linear modeling of complex data by artificial neural networks. Curr. Opin. Biotechnol. **13**(1), 72–76 (2002)
17. Kalogirou, S.A.: Artificial neural networks in renewable energy systems applications: a review. Renew. Sustain. Energy Rev. **5**(4), 373–401 (2001)
18. Kalogirou, S.A.: Applications of artificial neural-networks for energy systems. Appl. Energy. **67**(1–2), 17–35 (2000)
19. González-Figueredo, C., Rico-Martínez, R.: Towards bifurcation detection in kinetic monte carlo simulations: Robust identification with artificial neural networks and nonlinear Kalman filters. Int. J. Chem. React. Eng. **3**(1) (2006)
20. Gonzaga, J., et al.: ANN-based soft-sensor for real-time process monitoring and control of an industrial polymerization process. Comput. Chem. Eng. **33**(1), 43–49 (2009)
21. Hou, R., et al.: Unstructured big data analysis algorithm and simulation of internet of things based on machine learning. Neural Comput. Applic. **32**(10), 5399–5407 (2020)
22. Adnan, K., Akbar, R.: Limitations of information extraction methods and techniques for heterogeneous unstructured big data. Int. J. Eng. Bus. Manag. **11**, 1847979019890771 (2019)
23. Hasan, A.T., et al.: An adaptive-learning algorithm to solve the inverse kinematics problem of a 6 DOF serial robot manipulator. Adv. Eng. Softw. **37**(7), 432–438 (2006)
24. Khatti, J., Grover, K.S.: CBR prediction of pavement materials in unsoaked condition using LSSVM, LSTM-RNN, and ANN approaches. Int. J. Pavem. Res. Technol., 1–37 (2023)
25. Patel G.C., M., Krishna, P., Parappagoudar, M.B.: An intelligent system for squeeze casting process—soft computing based approach. Int. J. Adv. Manuf. Technol. **86**, 3051–3065 (2016)
26. Rangaswamy, H., et al.: Experimental analysis and prediction of strength of adhesive-bonded single-lap composite joints: Taguchi and artificial neural network approaches. SN Appl. Sci. **2**, 1–15 (2020)
27. Sibalija, T.V., et al.: A soft computing-based study on WEDM optimization in processing Inconel 625. Neural Comput. Applic. **33**(18), 11985–12006 (2021)
28. Benedetti, M., et al.: Energy consumption control automation using artificial neural networks and adaptive algorithms: proposal of a new methodology and case study. Appl. Energy. **165**, 60–71 (2016)
29. Alam, F., et al.: Analysis of eight data mining algorithms for smarter internet of things (IoT). Proc. Comput. Sci. **98**, 437–442 (2016)
30. Bolón-Canedo, V., et al.: Scalability analysis of ANN training algorithms with feature selection. In: Conference of the Spanish Association for Artificial Intelligence. Springer (2011)
31. Peteiro-Barral, D., et al.: Toward the scalability of neural networks through feature selection. Expert Syst. Appl. **40**(8), 2807–2816 (2013)
32. Rajaraman, V.: JohnMcCarthy—father of artificial intelligence. Resonance. **19**, 198–207 (2014)
33. Haykin, S., Network, N.: A comprehensive foundation. Neural Netw. **2004**(2), 41 (2004)

34. Zurada, J.M., Marks, I., Robinson, C.J.: Computational Intelligence Imitating Life. Institute of Electrical and Electronics, New York, NY (1994)
35. Ajith, A.: Artificial Neural Networks Handbook of Measuring System Design. John wiley & sons (2005)
36. Panchal, G., et al.: Behaviour analysis of multilayer perceptrons with multiple hidden neurons and hidden layers. Int. J. Comput. Theory Eng. **3**(2), 332–337 (2011)
37. Cao, W., et al.: A review on neural networks with random weights. Neurocomputing. **275**, 278–287 (2018)
38. Shettigar, A.K., et al.: Artificial bee colony, genetic, back propagation and recurrent neural networks for developing intelligent system of turning process. SN Appl. Sci. **2**(4), 660 (2020)
39. Patel G.C., M., et al.: Back propagation genetic and recurrent neural network applications in modelling and analysis of squeeze casting process. Appl. Soft Comput. **59**, 418–437 (2017)
40. Manjunath Patel, G.C., Shettigar, A.K., Parappagoudar, M.B.: A systematic approach to model and optimize wear behaviour of castings produced by squeeze casting process. J. Manuf. Process. **32**, 199–212 (2018)
41. Laje, R., Buonomano, D.V.: Robust timing and motor patterns by taming chaos in recurrent neural networks. Nat. Neurosci. **16**(7), 925–933 (2013)
42. Knecht, W.: Diesel engine development in view of reduced emission standards. Energy. **33**(2), 264–271 (2008)
43. Lloyd, A.C., Cackette, T.A.: Diesel engines: environmental impact and control. J. Air Waste Manag. Assoc. **51**(6), 809–847 (2001)
44. Iodice, P., Cardone, M.: Ethanol/gasoline blends as alternative fuel in last generation spark-ignition engines: a review on CO and HC engine out emissions. Energies. **14**(13), 4034 (2021)
45. Alrazen, H.A., Ahmad, K.: HCNG fueled spark-ignition (SI) engine with its effects on performance and emissions. Renew. Sustain. Energy Rev. **82**, 324–342 (2018)
46. Ramesh, T., et al.: Combined effect of compression ratio and fuel injection pressure on CI engine equipped with CRDi system using prosopis juliflora methyl ester/diesel blends. Int. J. Chem. Eng. **2022**, 1–12 (2022)
47. Aldhaidhawi, M., et al.: Ignition delay, combustion and emission characteristics of Diesel engine fueled with rapeseed biodiesel–A literature review. Renew. Sustain. Energy Rev. **73**, 178–186 (2017)
48. Fu, J., et al.: Application of artificial neural network to forecast engine performance and emissions of a spark ignition engine. Appl. Therm. Eng. **201**, 117749 (2022)
49. Zhou, Y., Gao, F.: Smart batch process: the evolution from 1D and 2D to new 3D perspectives in the era of big data. J. Process Control. **130**, 103088 (2023)
50. Rajkumar, S., Das, A., Thangaraja, J.: Integration of artificial neural network, multi-objective genetic algorithm and phenomenological combustion modelling for effective operation of biodiesel blends in an automotive engine. Energy. **239**, 121889 (2022)
51. Wong, K.I., et al.: Modeling and optimization of biodiesel engine performance using advanced machine learning methods. Energy. **55**, 519–528 (2013)
52. Aghbashlo, M., et al.: Machine learning technology in biodiesel research: a review. Prog. Energy Combust. Sci. **85**, 100904 (2021)
53. Vasudeva, S.T., et al.: Diagnostic classification of undifferentiated fevers using artificial neural network. In: AIP Conference Proceedings. AIP Publishing (2020)
54. Khurana, S., et al.: Predictive modeling of engine emissions using machine learning: a review. Mater. Today Proc. **38**, 280–284 (2021)
55. Climent, H., et al.: Analysis on the potential of EGR strategy to reduce fuel consumption in hybrid powertrains based on advanced gasoline engines under simulated driving cycle conditions. Energy Convers. Manag. **266**, 115830 (2022)
56. Kurani, A., et al.: A comprehensive comparative study of artificial neural network (ANN) and support vector machines (SVM) on stock forecasting. Ann. Data Sci. **10**(1), 183–208 (2023)
57. Grzegorzek, W., Ścieszka, S.: Prediction on friction characteristics of mine hoist disc brakes using artificial neural networks. Sci. Probl. Mach. Operat. Maint. ZEM. **46**, 27–42 (2011)

58. Deng, F., et al.: Magnetotelluric deep learning forward modeling and its application in inversion. Remote Sens. **15**(14), 3667 (2023)
59. Osman, O., Albora, A.M., Ucan, O.N.: Forward modeling with forced neural networks for gravity anomaly profile. Math. Geol. **39**, 593–605 (2007)
60. Sidrane, C., et al., Verifying inverse model neural networks. arXiv preprint arXiv:.02429 (2022)
61. LeCun, Y., Bengio, Y.: Convolutional networks for images, speech, and time series. Handbook Brain Theory Neural Netw. **3361**(10), 1995 (1995)
62. Li, X., Mao, W., Jiang, W.: Extreme learning machine based transfer learning for data classification. Neurocomputing. **174**, 203–210 (2016)
63. Pradhan, M.K., Minz, S., Shrivastava, V.K.: A kernel-based extreme learning machine framework for classification of hyperspectral images using active learning. J. Indian Soc. Rem. Sens. **47**, 1693–1705 (2019)
64. Vasudeva, S.T., et al.: Development of a convolutional neural network model to predict coronary artery disease based on single-lead and twelve-lead ECG signals. Appl. Sci. **12**(15), 7711 (2022)
65. Hellström, T., Dignum, V., Bensch, S.: Bias in machine learning—what is it good for? arXiv preprint arXiv:.00686 (2020)
66. Mehrabi, N., et al.: A survey on bias and fairness in machine learning. ACM Comput. Surv. **54**(6), 1–35 (2021)
67. Perez, A., Larranaga, P., Inza, I.: Supervised classification with conditional Gaussian networks: increasing the structure complexity from naive Bayes. Int. J. Approx. Reason. **43**(1), 1–25 (2006)
68. Bouthillier, X., et al.: Accounting for variance in machine learning benchmarks. Proc. Mach. Learn. Syst. **3**, 747–769 (2021)
69. Dar, Y., Muthukumar, V., Baraniuk, R.G.: A farewell to the bias-variance tradeoff? an overview of the theory of overparameterized machine learning. arXiv preprint arXiv:.02355 (2021)
70. Kantardzic, M.: Data Mining: Concepts, Models, Methods, and Algorithms. John Wiley & Sons (2011)
71. Eren, B., Guvenc, M.A., Mistikoglu, S.: Artificial intelligence applications for friction stir welding: a review. Metals Mater. Int. **27**, 193–219 (2021)
72. Tong, D.L., Mintram, R.: Genetic Algorithm-Neural Network (GANN): a study of neural network activation functions and depth of genetic algorithm search applied to feature selection. Int. J. Mach. Learn. Cybern. **1**, 75–87 (2010)
73. Stastny, J., Skorpil, V.: Genetic algorithm and neural network. In Proceedings of the 7th WSEAS International Conference on Applied Informatics and Communications. Citeseer (2007)
74. Katoch, S., Chauhan, S.S., Kumar, V.: A review on genetic algorithm: past, present, and future. Multimed. Tools Appl. **80**, 8091–8126 (2021)
75. Varun Kumar, S., Panneerselvam, R.: A study of crossover operators for genetic algorithms to solve VRP and its variants and new sinusoidal motion crossover operator. Int. J. Comput. Intell. Res. **13**(7), 1717–1733 (2017)
76. Lim, S.M., et al.: Crossover and mutation operators of genetic algorithms. Int. J. Mach. Learn. Comput. **7**(1), 9–12 (2017)
77. Mahajan, R., Kaur, G.: Neural networks using genetic algorithms. Int. J. Comput. Appl. **77**(14) (2013)
78. Parisi, G.I., et al.: Continual lifelong learning with neural networks: a review. Neural Netw. **113**, 54–71 (2019)
79. Filippi-Mazzola, E., Wit, E.C.: Modeling non-linear effects with neural networks in relational event models. arXiv preprint arXiv: 12357 (2023)
80. Patan, K., Patan, M.: Neural-network-based iterative learning control of nonlinear systems. ISA Trans. **98**, 445–453 (2020)
81. Sovány, T., et al.: Estimation of design space for an extrusion–spheronization process using response surface methodology and artificial neural network modelling. Eur. J. Pharm. Biopharm. **106**, 79–87 (2016)

82. Daviran, M., et al.: A new strategy for spatial predictive mapping of mineral prospectivity: automated hyperparameter tuning of random forest approach. Comput. Geosci. **148**, 104688 (2021)
83. Jagadish, et al.: Abrasive water jet machining for a high-quality green composite: the soft computing strategy for modeling and optimization. J. Braz. Soc. Mech. Sci. **44**(3), 83 (2022)
84. Manjunath Patel, G.C., et al.: Intelligent modelling of hard materials machining. In: Machining of Hard Materials: A Comprehensive Approach to Experimentation, Modeling Optimization, pp. 73–102. Springer (2020)
85. Jeong, C., Min, J.H., Kim, M.S.: A tuning method for the architecture of neural network models incorporating GAM and GA as applied to bankruptcy prediction. Expert Syst. Appl. **39**(3), 3650–3658 (2012)

Chapter 5
Optimization of Engine Performance and Emission Characteristics

5.1 Introduction

The transesterification process derived biodiesel from the Garcinia gummi-gutta seeds grown on non-arable forestland ensures low cost. The biodiesel blends were prepared with a combination of diesel and biodiesel. Experiments were designed to plan and conduct experiments, analyse the process insights, and derive empirical equations useful for prediction and optimization. The algorithms' performance relies on the algorithmic ability to converge to global solutions by balancing the exploration and exploitation in the search space, tuning of algorithm parameters, and computation time. A universal method has yet to be established in selecting the algorithm for conducting optimization tasks. Therefore, six metaheuristic algorithms were applied to enhance the performance and emission characteristics of diesel engines fuelled with biodiesel blends to diesel engines. The empirical equations derived through experimental investigations were used as objective functions to conduct multiple objective optimization tasks. A parametric study is conducted on all the algorithms to determine their effectiveness and improve solution accuracy. Objective functions are formulated using appropriate mathematical formulations, weight fractions assigned to individual outputs, and desirability function approaches to optimize the variables. The optimal variables for better performance and emission characteristics were evaluated by confirming experimental trials. These optimized conditions are further tested by adding silver oxide nanoparticles to the fuel blends. Finally, confirmation experiments are conducted to verify the results, ensuring that the optimal conditions are of practical relevance. The framework illustrating the metaheuristic algorithms optimizing the engine performance and emission characteristics is presented in Fig. 5.1.

M. Patel G. C. et al., *Biofuel Production, Performance, and Emission Optimization*, Springer Optimization and Its Applications 215,
https://doi.org/10.1007/978-3-031-84806-3_5

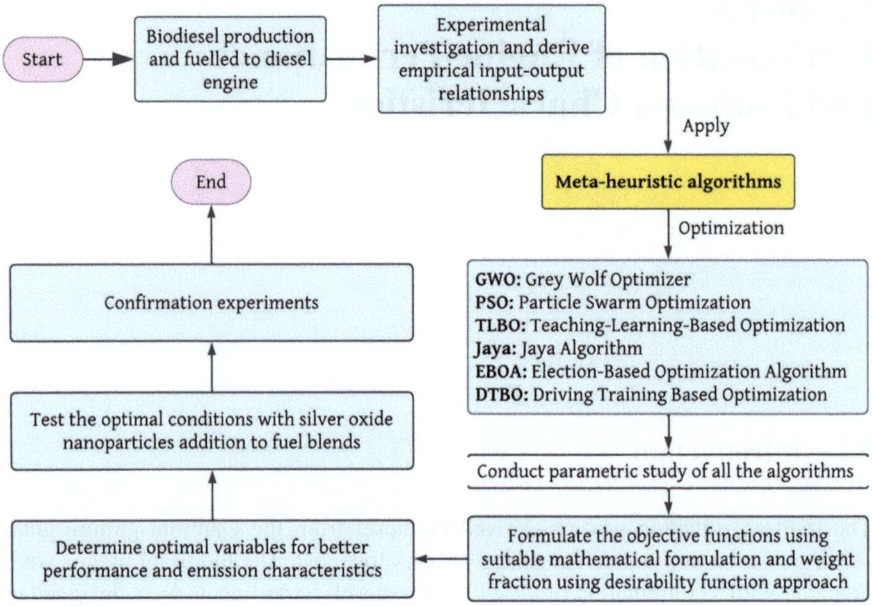

Fig. 5.1 Metaheuristic algorithms for optimizing input-output variables

5.2 Introduction to Selection and Working of Optimization Algorithms

Optimization techniques are typically employed to determine the best or optimal solution for a given problem by either maximizing or minimizing the provided characteristic or functions [1]. Numerous algorithms are highly beneficial in numerous scientific and technological domains. Some of the notable ones are the following: JAYA Algorithm, Grey Wolf Optimization (GWO), Particle Swarm Optimization (PSO), Teaching–Learning-Based Algorithm (TLBO), Election-Based Optimization Algorithm (EBOA), and Driving Training-Based Optimization (DTBO). These algorithms are not constrained by the characteristics of data sets, and they can be used in any scientific or technical field [2]. Currently, there are numerous optimization techniques that are equally successful within every application category or combination of features. In this part, various metaheuristic optimization methods and their operation are explained. These algorithms are used to optimize engine performance and emission characteristics.

5.2.1 GWO

The GWO algorithm was first postulated by Mirjalili and colleagues in 2014. The algorithm mimics the predatory strategies and hierarchical organization found in grey wolves [3]. Grey wolves are top predators that live in groups and adhere to a rigid social hierarchy consisting of four layers: alpha (α), beta (β), delta (δ), and omega (ω) [4]. The most powerful individuals and decision-makers are the alphas. The second level is made up of betas that help alphas make decisions. When the alpha wolves pass away or age, they could step in as a replacement. In comparison to omega wolves, which alter their positions in respect to α, β, and δ wolves, delta wolves are more dominant, as stated by Mirjalili. The following are included in the framework of the GWO algorithm steps [3, 4].

Initialize population, algorithm settings, and wolf positions $[x_i = 1, \ldots w]$ between the boundaries. The wolves' (w) initialization at random sets the stage for prey searching and detection.

Step.1. Determine the OF for each wolf, noting that hunting is predominantly led by the α, β, and δ wolves, with the most effective solutions being associated first with α, then β, and finally δ. The last three answers deal with the wolves, who follow the route taken by the first three groups.

Step.2. The wolf's location during the encirclement of prey is determined by its own position as well as the vectors $\vec{B}$ and $\vec{C}$, which aid in the formation of hyperspheres by wolves (refer to Eq. 5.1 and 5.2). The calculation is based on the vector $\vec{a}$ along with two random vectors, $\vec{r_1}$ and $\vec{r_2}$ each having values ranging from 0 to 1.

$$\vec{B} = 2\vec{a} \cdot \vec{r_1} - \vec{a} \tag{5.1}$$

$$\vec{C} = 2\vec{r_2} \tag{5.2}$$

The vector $\vec{a}$ gradually reduces from two to zero in a linear fashion, based on the current count of iterations relative to the overall number of iterations. This feature improves the exploration vs. exploitation equilibrium.

Step.3. During the prey hunt, the positions of the wolves are adjusted by moving them within a continuous space around the prey. The ideal location is determined by averaging the best solutions provided by the α, β, and δ wolves.

Step.4. At last, the method will verify the termination requirement. If criterion (i) is satisfied, then consider the α wolf position to be the ideal one and end the process; (ii) if not, begin the process anew at step 2.

5.2.2 PSO

The PSO simulates how insects forage, i.e. how a swarm of particles moves in a complex environment at multiple locations at once to obtain a global position for both an individual and a swarm [5]. This method will solve the complicated optimization issues more quickly since it just needs a small number of well-understood tuning parameters (acceleration coefficients, inertia weight (ω)) to operate [6]. Key PSO parameters, such asω, swarm size, and the maximum number of iterations, serve as tuning variables for optimization. The PSO algorithm's specific steps are listed below.

1. Initialize PSO: Considering the top and lower bounds for n design variables, create a swarm with locations $X_i(i = 1, 2, 3\ldots\ldots$Particle) and velocities S_i.
2. For each x_i, evaluate the objective function $f(x_i)$ determine the optimal location for each particle P_{best}, as well as the overall optimal location G_{best} across the population, including their respective objective values.
3. Adjust the velocities in accordance with Eq. (5.3). To allow the particles to travel while preventing an early convergence, the ω regulates the velocity. The particle cognition and social regulatory elements D_1 and D_2 that helps in avoiding an early convergence. The random vectors, viz., u_1 and u_2, whose lengths match the number of variables, with values that range from 0 to 1.

$$S_i^{(k+1)} = w \cdot S_i^k + D_1 u_1 \left(P_{\text{best}} - x_i^k\right) + D_2 u_2 \left(G_{\text{best}} - x_i^k\right) \tag{5.3}$$

4. Now $x_i^{(k+1)}$ location is updated using the below Eq. 5.4:

$$x_i^{(k+1)} = x_i^k + D_i^{(k+1)} \tag{5.4}$$

5. Evaluate the objective functions (OF) for updating solutions to identify new $P_{\text{best}}$and G_{best}, and subsequently refresh the swarm.
6. Examine the termination criteria: end the process if they are met; otherwise, restart the process from step 2 if they are not satisfied.

Optimizing the specific parameters of the algorithm for each application is essential for attaining the best performance.

5.2.3 TLBO

The TLBO technique was established by Rao et al. [7] based on the idea that teachers have an impact on the students in their class. TLBOs imitate how teachers and students can teach and learn in a classroom [8]. The TLBO algorithm was created to reduce the drawbacks associated with adjusting algorithm-specific parameters (GA necessitates adjusting crossover and mutation rate, while PSO requires adjusting ω)

[9]. As of yet, no accepted global value or set of standards has been established to identify the parameters unique to each algorithm [10, 11]. These values are particular to the issue, and incorrect adjustment of algorithm-specific parameters may lead to longer calculation times and local solutions [12, 13].

TLBO functions in two stages: the teacher phase and the learner phase [14–16]. In TLBO, the various subjects that a set of learners is exposed to are considered design variables, and the learners' results define the problem domain's fitness value. The learners are viewed as a population. The main duties of the teacher and learner phases in the TLBO process are examined in the section that follows.

Teacher phase: To the best of their knowledge, a teacher's primary goal when teaching a class is to raise the average student performance in the material they are covering. The mathematical techniques utilized to identify the top student, determine the teaching factor (TF), and evaluate the variation in the solution that was updated during the teacher phase have all been fully explained in earlier work [12, 17]. After the instructor phase is over, the revised and approved solutions are saved and used as input for the learner phase.

Learner phase: By interacting with other students at random, students attempt to expand on the knowledge they have already learned from the teacher. During this stage, students might pick up new knowledge while also having the chance to learn from someone who is more knowledgeable or skilled than them. The literature [12, 17] provides an explanation of the mathematical steps that are engaged in the learner phase. For the following iteration, the approved and revised learner phase solutions are used as input in the instructor phase. Until the predetermined termination requirements are satisfied, the teacher and learner phases are repeated.

5.2.4 JAYA Algorithm

JAYA, which means "victory" in Sanskrit, is the name of another algorithm-specific parameter-less algorithm that Rao suggested in 2016 [18]. The JAYA algorithm is predicated on a straightforward, widely recognized idea: the best (global) solutions to the problem are approached from a distance, and the worst (local) answers are avoided [18, 19]. Equation (5.5) is utilized to replace the population.

$$
\begin{aligned}
&x'_{(i,k,l)} = x_{(i,k,l)} + rand_{(1,i,l)}[0,1]\left(x_{(i,\text{best},l)} - \|x_{(i,k,l)}\|\right) \\
&-rand_{(2,i,l)}[0,1]\left(x_{(i,\text{worst},l)} - \|x_{(i,k,l)}\|\right) \\
&x_{(i,k,l)} : i\text{th variable}\left(n\,\text{design variables}\right) \\
&of\ k\text{th population}\left(m\,\text{members}\right)\text{ in the } l\text{th iteration;} \\
&x_{(i,\text{best},l)} \text{ and } x_{(i,\text{worst},l)} : \text{values of } i\text{th variable} \\
&\text{of the best and worst members of population in } l\text{th iteration}
\end{aligned}
\tag{5.5}
$$

Where $\text{rand}_{(1,i,l)}[0, 1]$ and $\text{rand}_{(2,i,l)}[0, 1]$ used to advance the exploration capacity.

The members' tendency to migrate towards favourable solutions is reflected in the part of the equation that contains rand 1, *k*, *l*, while their tendency to move away from the worst solutions is reflected in the part of the equation that includes rand 2, *k*, *l*.

Next, the modified solution $x'_{(i,k,l)}$ is evaluated with respect to the fitness function *f*(*x*). If the adjusted solution result in an improved OF value, it will be integrated into the next iteration.

5.2.5 EBOA

This algorithm functions on the basis of an election-voting process, where an election is defined as a procedure by which community members choose or cast ballots for a candidate [20]. The person elected as a leader influences the conditions of every member in the society, including those who did not vote for them. Being better informed enables community members to more effectively choose and vote for the most deserving candidate [21].

Initially, this process will create a population matrix [X] based on EBOA membership, and then use Eq. (5.6) to randomly identify each person's starting position in the search space.

$$x_{i,j} = \underbrace{\mathrm{lb}_j}_{\substack{\text{lower bound of}\\ j\text{th variable}}} + rand[0,1] \bullet \left(\underbrace{\mathrm{ub}_j}_{\substack{\text{upper bound of}\\ j\text{th variable}}} - \mathrm{lb}_j \right), \ldots\ldots\ldots\ldots i = 1,2,\ldots\ldots\ldots m \quad (5.6)$$

Based on each EBOA member's recommendations for the problem variables, the OF value can be ascertained. These evaluated values for the OF of the issue are given using a vector that conforms with Eq. (5.7).

$$\underbrace{\mathrm{OF}}_{\text{Objective function}} = \begin{bmatrix} \mathrm{OF}_1 \\ \vdots \\ \mathrm{OF}_2 \\ \vdots \\ \vdots \\ \mathrm{OF}_N \end{bmatrix}_{N\times 1} = \begin{bmatrix} \mathrm{OF}(X_1) \\ \vdots \\ \mathrm{OF}(X_2) \\ \vdots \\ \vdots \\ \mathrm{OF}(X_N) \end{bmatrix} \quad (5.7)$$

Where OF_i denotes the *i*th EBOA member's OF value and OF stands for the vector of the EBOA population's obtained OF values. The OF values act as a benchmark by which to evaluate the quality of proposed solutions; the member with the highest function value is considered to be the best, and the member with the lowest value is considered to be the worst.

5.2.6 *Mathematical Model of EBOA [20]*

The subject matter addresses the first two phases of the algorithm population update process in EBOA, which are exploration and exploitation.

Phase of Exploration: Elections and voting procedures [20].

Members of EBOA take part in the election and cast a ballot for a candidate based on their level of awareness. It could be argued that people's awareness is shaped by the quality and perceived merit of the OF value. Consequently, the application of Eq. (5.8) mimics the knowledge of community members. Increased awareness is indicated by better OF scores in this awareness simulation process.

$$\underbrace{A_i}_{\text{Awareness of the } i\text{th EBOA member}} = \begin{cases} \dfrac{\text{OF}_i - \text{OF}_{\text{worst}}}{\text{OF}_{\text{best}} - \text{OF}_{\text{worst}}}, & \text{OF}_{\text{best}} \neq \text{OF}_{\text{worst}}; \\ 1, & \text{else}, \end{cases} \tag{5.8}$$

OF_{best} and OF_{worst} : best and worst values of OF

10% of the public's most aware members are believed to be qualified candidates for public office. The EBOA predicts that at least two candidates (NC = 2; i.e. NC $\geq$ 2) will file to run in the election.

To cast an EBOA vote, one must compare each voter's awareness level to a number that is created at random. Voting for the front-runner or best candidate, C1, is open to everyone whose awareness level is higher than this random number. A voter will select one of the other candidates at random if their awareness level is not greater than the number that is created at random. A mathematical model of this voting process is provided by Eq. (5.9).

$$\underbrace{v_i}_{\text{Vote of } i\text{th person in the community}} = \begin{cases} C_1 \\ C_k, \end{cases} = \begin{cases} \dfrac{\text{Best candidate}}{\text{randomly selected } k\text{th candidate from set of } 2,3,\ldots \text{Nc}}, \end{cases} \quad \begin{matrix} A_i > r; \\ \text{else} \end{matrix} \tag{5.9}$$

The candidate who receives the most votes is declared the winner (leader). Every member of society is impacted by the decisions made by the elected leader, even those who did not cast ballots for them. The EBOA is updated for each individual position in accordance with directives from the elected leader. Enhancing the EBOA's global search capabilities, the leader directs the population of the algorithm to explore different regions of the search space. The leader initiates the update process within the EBOA population by generating a new position for each member.

Update it if the newly created position raises the OF's value. The matching member returns to their previous location if not. The EBOA updating process is characterized by using Eqs. (5.10) and (5.11).

$$\underbrace{x_{i,j}^{\text{new}P1}}_{\text{Newly generated position for } i\text{th EBOA member}} = \begin{cases} x_{i,j} + rand[0,1].\left(L_j - I[1 \text{or} 2].x_{i,j}\right), \text{OF}_L < \text{OF}_i; \\ x_{i,j} + rand[0,1].\left(x_{i,j} - \underbrace{L_j}_{\text{Elected leader of } j\text{th dimension}} \right), \text{else}, \end{cases} \tag{5.10}$$

$$X_i = \begin{cases} X_i^{\text{new}.P1} \text{OF}_i^{\text{new},P1} < \text{OF}_i; \\ X_{i,} \text{else} \end{cases} \tag{5.11}$$

Where L refers to the leader who was chosen and OFL stands for the leader's OF value.

Phase 2 of exploitation: public awareness-raising campaign [20].

The level of awareness among society's members greatly influences their ability to make wise choices during the election and voting process. Apart from the impact of the leader on individuals' consciousness, each person's ideas and actions have the power to heighten their own awareness. Mathematically speaking, a local search close to any recommended solution might offer a better choice. Therefore, the EBOA's ability to use local searches and find better solutions to problems is enhanced by community members' efforts to raise awareness [20].

In the search space, a random point near each member is taken into consideration to mimic this local search process. Then, in order to determine whether the OF of the issue is better than the member's current situation, it is evaluated in the context of the new situation. The local search is considered successful if the new position produced a higher value for the OF, in which case the relevant member's position is modified. Enhancing the OF value will make that person more conscious, enabling them to make wiser decisions in the upcoming election (or iteration). Eqs. (5.12) and (5.13) provide a framework for enhancing public awareness of the EBOA during this update process.

$$\underbrace{X_{i,j}^{\text{new}P2}}_{\text{Updated position for } i\text{th EBOA member}} = \left\{ x_{i,j} + (1-2r).R[0.2].\left(1 - \frac{t(\text{iteration contour})}{T(\text{Maximum number of iteration})}\right).x_{i,j} \right. \tag{5.12}$$

$$X_i = \begin{cases} X_i^{\text{new}.P2} \text{OF}_i^{\text{new},P2} < \text{OF}_i; \\ X_{i,} \text{else} \end{cases} \tag{5.13}$$

Where $X_i^{\text{new}.\,P2}$ is its jth dimension, $\text{OF}_i^{\text{new},\,P2}$ is its value of the OF.

5.2.7 DTBO

DTBO, an automated optimization process, trains beginners in driving skills, allowing learner drivers to choose any instructor at the outset when joining a driving school [22]. Subsequently, the instructor imparts instructions and skills to the learner driver. The learner driver strives to learn driving techniques from the instructor and follows them while driving [23]. Furthermore, individual practice helps enhance the learner's driving abilities. The potential for designing an optimizer through these interactions and activities is great. This explanation explains how the mathematical modelling of the process served as a major inspiration for the design of DTBO.

Initially, a population matrix made up of instructors and driving learners was created to generate DTBO. The issue variables are given values by each possible solution and then evaluated for the OF by adding them in, as stated by Eq. (5.14). To determine a possible solution, an OF value is therefore calculated.

$$\mathrm{OF} = \begin{bmatrix} \mathrm{OF}_1 \\ \vdots \\ \mathrm{OF}_2 \\ \vdots \\ \vdots \\ \mathrm{OF}_N \end{bmatrix}_{\underbrace{N}_{\text{Size of population}} \times 1} = \begin{bmatrix} \mathrm{OF}(X_1) \\ \vdots \\ \mathrm{OF}(X_2) \\ \vdots \\ \vdots \\ OF(X_N) \end{bmatrix}_{N \times 1} \tag{5.14}$$

Where X_i is the ith candidate solution of DTBO.

The values produced for the OF are the main metric used to assess candidate solutions' quality. After comparisons, the member with the highest OF value is regarded as the best in the population (X_{best}). The best member has to be updated as candidate solutions improve with each cycle.

The DTBO method works in three phases:

- *Exploration Phase: Training by the driving instructor* [22]
  A driving instructor (DI) is selected by the trainee driver during the first phase of the DTBO update, after which they take driving lessons from them. Each NDI member iteration is selected as an instructor in this phase according to their OF values, as indicated by Eq. (5.15) [22].

$$\underbrace{\text{DI}}_{\text{Matrix of DI}} = \begin{bmatrix} \text{DI}_1 \\ \vdots \\ \vdots \\ \text{DI}_i \\ \vdots \\ \vdots \\ \text{DI}_{NDI} \end{bmatrix}_{N_{\text{DI}} \times m} = \begin{bmatrix} \text{DI}_1 & \cdots & \text{DI}_{1j} & \cdots & \text{DI}_{1m} \\ \vdots & \vdots & \vdots & \vdots & \vdots \\ \text{DI}_{i1} & \cdots & \text{DI}_{ij} & \cdots & \text{DI}_{im} \\ \vdots & \vdots & \vdots & \vdots & \vdots \\ \text{DI}_{NDI1} & \cdots & \text{DI}_{NDI1j} & \cdots & \text{DI}_{NDIm} \end{bmatrix}_{NDI \times m} \quad (5.15)$$

Where DI_i is the ith DI, $\text{DI}_{i,j}$ is the jth dimension, and $N\text{DI} = [0.1 \cdot N \cdot (1 - t/T)]$ is the number of DI.

This DTBO phase's mathematical modelling is set up so that, at first, Eq. (5.16) is used to determine each member's new position. The old position is then replaced if this new position raises the value of the OF, as specified in Eq. (5.17). However, in order for DI to train the ith member, $F\text{DI}_{k\,i}$ is its OF value, and $\text{DI}_{k\,i,j}$ is its jth dimension.

$$\underbrace{x_{i,j}^{P1}}_{\substack{\text{Updated } i\text{th candidate} \\ \text{solution from } j\text{th dimension}}} = \begin{cases} x_{i,j} + rand[0,1].\left(\text{DI}_{k_i}, j - I[1,2].x_{i,j}\right), F_{\text{DI}_{Ki}} < F_i; \\ x_{i,j} + E[0,1].\left(x_{i,j} - \text{DI}_{k_i}, j\right), \text{otherwise}, \end{cases} \quad (5.16)$$

$$X_i = \begin{cases} X_i^{P1} F_i^{P1} < F_i; \\ X_{i,} \text{otherwise} \end{cases} \quad (5.17)$$

Where $F^{P1}{}_i$ is its OF value, DI_{ki}, where randomly selected$K_i = \{1, 2, \ldots, N_{DI}\}$.

- *Exploration Phase: Patterning of the instructor skills of the student driver* [22] In the second phase of the DTBO update, the student driver imitates all of the instructor's movements and skills. This procedure moves the members of the DTBO to different locations around the search space, improving its exploratory power. To mathematically illustrate this idea, we apply Eq. (5.18) to build a new position across the linear set of each member with the teacher. The new location takes precedence over the previous one if it increases the OF value, as stated in Eq. (5.19).

$$x_{i,j}^{P2} = P.x_{i,j} + (1 - P).\text{DI}k_{i,j} \quad (5.18)$$

$$X_i = \begin{cases} X_i^{P2} F_i^{P2} < F_i; \\ X_{i,} \text{otherwise} \end{cases} \quad (5.19)$$

Where X_i^{P2} is the updated status of *i*th solution of candidate correspond to exploration phase of DTBO, $x_{i,j}^{P2}$ is its *j*th dimension, F_i^{P2} is its OF value. Eq. 5.20 presents the computation of patterning index.

$$\underbrace{P}_{\substack{\text{Patterning}\\ \text{index}}} = 0.01 + 0.9\left(1 - \frac{t}{T}\right) \tag{5.20}$$

- *Exploitation Phase: Personal practice* [22]
  In the third phase of the DTBO update, each trainee driver's individual practice is the main focus, with an emphasis on skill growth and improvement. Every student driver seeks to maximize their driving abilities throughout this phase. Each member has the opportunity to discover an improved position through a local search cantered around their current location. This phase demonstrates DTBO's ability to use local search efficiently. According to Eq. (5.21), each member of the population must initially be placed in close proximity to a random point. In the event when the new position uses Eq. (5.22) to increase the OF value, the old one is substituted.

$$x_{i,j}^{P3} = x_{i,j} + \left(1 - 2 \times rand\left[0,1\right]\right).R.\left(1 - \frac{t}{T}\right).x_{i,j} \tag{5.21}$$

$$X_i = \begin{cases} X_i^{P3} F_i^{P3} < F_i; \\ X_{i,}\text{otherwise} \end{cases} \tag{5.22}$$

5.3 Optimization of Biodiesel Yield

Optimization is the process of identifying the extreme values of the outputs' objective functions and the corresponding inputs [24]. The extreme values may maximize or minimize the objective functions subjected to input variable constraints [25]. The optimization task could help to determine the transesterification parameter values responsible for maximum biodiesel yield. The optimization effort involved applying six algorithms (GWO, PSO, TLBO, JAYA, DTBO, and EBOA) to discover the best spots responsible for maximal biodiesel conversion using the BBD-derived response equation as an objective function. Eq. (5.23) is used as an objective function, and their constrained limits of transesterification variables are discussed as follows:

$$\begin{aligned}\text{Yield}_{\text{BBD}} = {} & 259.7 - 40.75\text{A} + 1053.6\text{B} + 0.963\text{C} + 0.6391\text{A}^2 - 544.9\text{B}^2 \\ & - 0.00232\text{C}^2 - 2.5\text{AB} + 0.0037\text{AC} - 0.717\text{BC}\end{aligned} \tag{5.23}$$

Constraints

$$30 \leq \text{Methanol}(A) \leq 40$$
$$0.6 \leq \text{NAOH Concentration}(B) \leq 1.0$$
$$45 \leq \text{Reaction time}(C) \leq 75$$

The optimal condition that optimizes the biodiesel yield was found by all algorithms to be the same. The programme calculated the ideal parameters, which produced a maximum biodiesel conversion of 94.01% (methanol: 40 mL, reaction time: 75 min, sodium hydroxide concentration: 0.83 g). The resulting experimental biodiesel yield was found to be equal to 96.2%.

5.4 Need of Optimization of Engine Performance and Emission Characteristics

The practical usefulness of determining the extreme maximum values of BTE and the minimal values of BSFC, CO, NOx, and UHC lies in their ability to improve performance and emission characteristics in diesel engines. For the reasons listed in Table 5.1, it is difficult to identify a single input condition that satisfies the competing requirements (Minimize: BSFC, CO, NOx, and UHC, and Maximize: BTE) in outputs.

Table 5.1 displays how process factors affected every output that displayed conflict, with the exception of NOx. For instance, BT's maximum biodiesel % compromises with lower BTE and higher BSFC and NOx emissions in place of CO and UHC emissions. In addition to providing compromises with BSFC and NOx emissions, higher CR is beneficial for BTE, CO, and UHC. Investigators must therefore ascertain which single conditions (BT, EL, CR, and IP) in the objective output functions (BTE, NOx, UHC, CO, and BSFC) fulfil competing needs. The multi-objective optimization problem that resulted from this scenario can be summed up as follows:

Table 5.1 Summary results of factors influence on different outputs

Input variable	Performance characteristics	Emission characteristics	References
Increase in BT (10–30%)	↑ in BSFC, ↓ BTE	↑ in NOx, ↓ CO, UHC	Figs. 3.28, 3.30, 3.32, 3.34, and 3.36
Increase in CR (14–18)	↑ in BTE, ↓ BSFC	↑ in NOx, ↓ CO, UHC	
Increase in EL (20–100)	↑ in BTE, ↓ BSFC	↑ in NOx, UHC, ↑initially increase and ↓ CO	

1. Finding a single solution that meets all requirements is difficult when there are conflicting objectives (e.g. increasing BTE may result in increased NOx emissions). MOO handles complex decision-making procedures in these situations.
2. A paradigm for comprehending and evaluating the trade-offs between performance and emission targets is offered by MOO. Also ascertains which combinations of input factors (such as BT, IP, CR, and EL) result in the best possible balance between emissions and engine efficiency.
3. Finding the best circumstances resulted in balancing emissions and performance, where MOO helps achieve engine specifications sustainably and reduces the impact of diesel engines on the environment.
4. New techniques for fuel injection, combustion, and exhaust treatment that enhance engine performance and emissions may be motivated by MOO results.
5. Enhanced engine durability and dependability, as well as decreased fuel consumption and emissions-related expenses, are just a few of the financial advantages that can arise from optimizing engine performance and emission characteristics.

5.5 Multi-Objective Optimization

The MOO involves a decision-making process that finds the best possible solutions to problems with multiple conflicting objectives. These objectives often conflict with each other, meaning that improving one objective may lead to the deterioration of another. The essence of the multi-objective optimization process lies in balancing these trade-offs to identify many optimal solutions (also referred to as pareto-optimal solutions). However, the investigators are more interested in finding a solution that satisfies all conflicting outputs.

MOO problems can be solved by traditional techniques such as linear programming or gradient descent, as well as artificial intelligence-based metaheuristic algorithms. Traditional techniques determining the best solution for a single-objective function in convex and well-defined problem spaces. However, they may struggle to find optimal solutions that balance competing objectives in complex scenarios. On the other hand, AI-based metaheuristic algorithms are designed to efficiently explore spaces with multiple search distinct or spatial locations. These algorithms drawn inspiration from natural phenomena, enabling them to optimize conflicting objective functions simultaneously and navigate through non-convex, discontinuous and large search spaces to identify a set of optimal solutions. The advantages of AI-based metaheuristic algorithms over traditional methods are discussed below:

- Adaptability to address different problems without major modification to the algorithm structure.
- Flexibility to tackle real-world problems where objectives and constraints may change over time.

- Ability to handle uncertainty, non-linearity, and multi-dimensionality to optimize complex problems and provide diverse solutions giving decision-makers suitable options based on their preferences and trade-offs.
- Ability to find near-optimal solutions within a reasonable timeframe.

5.6 Framework Illustrating the Steps in Multi-Objective Optimization

The steps in performing multi-objective optimization are illustrated in Fig. 5.2 and are explained by applying the following steps:

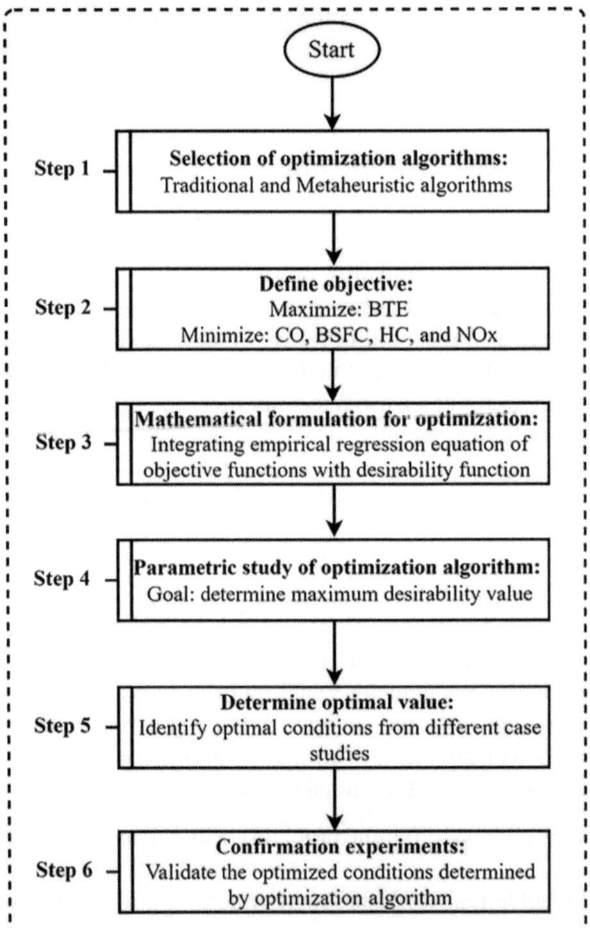

Fig. 5.2 Steps illustrating the framework for determining optimal conditions

5.6.1 Selection of Optimization Algorithms

Step 1 Many nature-inspired algorithms have recently been applied to solve optimization problems. Although all algorithms aim to determine global solutions to a problem under investigation, the working principles of each algorithm differ. Therefore, it requires tuning of algorithm-specific parameters to obtain global solutions. Table 5.2 summarizes algorithm working principles, control parameters, and problem application.

The performance of each algorithm differs when applied to solve a common problem. This is probably due to the requirement of tuning algorithm-specific parameters [7, 9]. Similarly, other algorithms, such as Grey Wolf Optimizer (GWO),

Table 5.2 Summary of algorithms applied for optimization problem

Algorithm	Working principle	Algorithm-specific parameters	Refs.
GA	Darwin's theory of survival of fittest	Crossover, mutation rate, selection operator	[26]
PSO	Foraging behaviour of bird searching for food	Inertia weight, social and cognitive leaders	[11]
ABC	Foraging behaviour of honey bee	Number of scouts, onlookers employed	[27]
ACO	Ant behaviour in searching the destination from source	α, β parameters influence pheromone, evaporation rate ρ	[28]
HS	Music improvisation in music players	Harmony memory consideration rate, pitch adjusting rate, and number of improvisations	[29]
GWO	Social hierarchy and hunting behaviour of grey wolves	Coefficient vectors such as A and C, and convergence parameter: a	[3]
DTBO	Process and experience of learning to drive in a driving school	Nil	[22]
EBOA	Voting process to select a leader	Nil	[20]
TLBO	Teaching-learning process in a class room	Teaching factor	[7]
JAYA	Approach of moving towards the best solution and away from the worst	Nil	[18]

Differential Evolution (DE), Bacteria Foraging Optimization (BFO), Firefly Algorithm (FA), and Big Bang–Big Crunch (BB-BC) algorithm, require tuning their respective algorithm-specific parameters [3, 30–33]. Inappropriate tuning of algorithm-specific parameters results in either local solutions or being trapped at local optimal solutions [18]. Unlike Genetic Algorithm (GA), Particle Swarm Optimization (PSO), Artificial Bee Colony (ABC), Ant Colony Optimization (ACO), Harmony Search (HS), DE, GWO, BB-BC, FA, and so on, TLBO, JAYA, DTBO, and EBOA do not require tuning algorithm-specific parameters DE, GWO, BB-BC, FA, and so on, TLBO, JAYA, DTBO, and EBOA do not require tuning algorithm-specific parameters [7]. The algorithms (such as DTBO, EBOA, and JAYA) require common control parameters (population size and maximum number of generations or iterations) based on their working principles, similar to other algorithms. To date, no specific algorithm has been accepted as a universal algorithm that could determine the best solutions for all problem domains. Therefore, the present work is focused on comparing all six algorithms (GWO, PSO, DTBO, EBOA, JAYA, and TLBO) that could solve multi-objective optimization problems related to biodiesel performance and emission characteristics.

5.6.1.1 Define Objective Function

Step 2 The optimization task starts with defining the objective function. The optimization task aims to minimize the BSFC, CO, NOx, and HC emissions and maximize the BTE.

5.6.1.2 Mathematical Formulation of the Problem

Step 3 In the present work, five outputs affect the engine performances and emission characteristics. The five outputs conflict with one another, i.e. maximizing the BTE and minimizing the BSFC, CO, NOx, and UHC. Solving multi-objective optimization requires transforming multiple objective functions into a single-objective function by employing weight fractions for maximization or minimization using suitable mathematical formulations. Equal weight methods and expert advice were adopted to determine the weight fractions. It is noteworthy that the solution accuracy varies with the importance given to individual objective functions. To date, there has yet to be a universal method established to determine the weight fractions for individual objective functions. Therefore, six case studies representing different sets of weights were adopted for the objective functions to solve multiple objective optimizations. The summation of a set of weights assigned to an individual function must be kept equal to 1. Six case studies with different sets of weights were presented in Table 5.3.

Terms W_1, W_2, W_3, W_4, and W_5 represent the weight fraction correspond to BTE, BSFC, CO, UHC, and NOx. The procedure employed to formulate different responses using weighted objective functions for maximization is shown below:

Table 5.3 Case studies with different sets of weights

Case study	Weight fraction assigning method	Weight fraction
Case 1	Equal weight fractions for all objective functions for all outputs	W_1 to $W_5 = 0.2$
Case 2	Maximum weight fraction to objective function for BTE	$W_1 = 0.6$, W_2 to $W_5 = 0.1$
Case 3	Maximum weight fraction to objective function for BSFC	$W_2 = 0.6$, W_1, W_3 to $W_5 = 0.1$
Case 4	Maximum weight fraction to objective function for CO	$W_3 = 0.6$, W_1, W_2, W_4, $W_5 = 0.1$
Case 5	Maximum weight fraction to objective function for UHC	$W_4 = 0.6$, W_1 to W_3, $W_5 = 0.1$
Case 6	Maximum weight fraction to objective function for NOx	$W_5 = 0.6$, W_1 to $W_4 = 0.1$

The objective functions representing different output functions are presented in Eqs. (5.24), (5.25), (5.26), (5.27), and (5.28):

$$\begin{aligned}
&\text{OF of BTE} = \\
&53.41375 - 0.006986\text{A} + 0.027977\text{B} - 4.21374\text{C} - 0.239568\text{D} \\
&+ 0.001038\text{A}^2 - 0.000058\text{B}^2 + 0.130046\text{C}^2 - 0.009348\text{D}^2 \\
&- 0.0000059375\text{AB} - 0.000445\text{AC} + 0.000098\text{AD} + 0.000044\text{BC} \\
&+ 0.000134\text{BD} + 0.025281\text{CD}
\end{aligned} \tag{5.24}$$

$$\begin{aligned}
&\text{OF of BSFC} = \\
&1.41019 - 0.002548\text{A} - 0.001998\text{B} - 0.09001\text{C} + 0.003605\text{D} \\
&- 0.00000229167\text{A}^2 + 0.00000373333\text{B}^2 + 0.002083\text{C}^2 \\
&- 0.000102\text{D}^2 + 0.0000003125\text{AB} + 0.000072\text{AC} \\
&+ 0.000021\text{AD} + 0.00001\text{BC} + 0.000003\text{BD} + 0.000294\text{CD}
\end{aligned} \tag{5.25}$$

$$\begin{aligned}
&\text{OF of CO} = \\
&- 0.331066 + 0.00589\text{A} + 0.001706\text{B} + 0.114789\text{C} - 0.052457\text{D} \\
&- 0.000021\text{A}^2 - 0.00000748148\text{B}^2 - 0.004676\text{C}^2 + 0.000713\text{D}^2 \\
&+ 0.00000125\text{AB} - 0.000141\text{AC} - 0.000019\text{AD} + 0.00005\text{BC} \\
&+ 0.000018\text{BD} + 0.000875\text{CD}
\end{aligned} \tag{5.26}$$

$$\begin{aligned}
&\text{OF of UHC} = \\
&161.81035 + 0.48066\text{A} - 0.042681\text{B} - 4.7789\text{C} - 6.83002\text{D} \\
&- 0.002477\text{A}^2 + 0.000015\text{B}^2 + 0.009259\text{C}^2 + 0.09037\text{D}^2 \\
&+ 0.000094\text{AB} - 0.000781\text{AC} - 0.000781\text{AD} \\
&- 0.000625\text{BC} + 0.000375\text{BD} + 0.121875\text{CD}
\end{aligned} \tag{5.27}$$

$$
\begin{aligned}
&\text{OF of NOx} = \\
&-403.66824 + 6.57076\text{A} - 0.369917\text{B} + 46.97569\text{C} \\
&-16.95181\text{D} + 0.005799\text{A}^2 - 0.004289\text{B}^2 - 1.80556\text{C}^2 \\
&+0.927778\text{D}^2 - 0.003812\text{AB} - 0.304687\text{AC} - 0.000312\text{AD} \\
&+0.15\text{BC} + 0.0215\text{BD} + 0.8625\text{CD}
\end{aligned}
\tag{5.28}
$$

Constraints

$$
\begin{aligned}
&20 \le \text{Engine load or } A \ge 100 \\
&160 \le \text{Injection pressure or } B \ge 260 \\
&14 \le \text{Compression ratio or } C \ge 18 \\
&10 \le \text{Blend type or } D \ge 30
\end{aligned}
$$

The mathematical formulation used to transform multiple objective functions into a single-objective function for maximization is presented in Eq. (5.29). The present work involves the conflicting nature of output functions. The desirability function approach is applied to transform the conflicting nature of multiple outputs to a single output function aiming towards maximization. The composite desirability (D_o) value, or fitness function value, is determined by considering all response functions using Eq. (5.29). The D_o value varies between zero and 1. The D_o value close to 1 signifies the ideal value for any process optimization.

$$
D_o = \sqrt[5]{\left(d_{\text{BTE}}^{W_1} \times d_{\text{BSFC}}^{W_2} \times d_{\text{CO}}^{W_3} \times d_{\text{UHC}}^{W_4} \times d_{\text{NOx}}^{W_5}\right)} \tag{5.29}
$$

$$
d_{\text{BTE}} = \frac{\text{BTE} - \text{BTE}_{\min}}{\text{BTE}_{\max} - \text{BTE}_{\min}} \tag{5.30}
$$

$$
d_{\text{BSFC}} = \frac{\text{BSFC}_{\max} - \text{BSFC}}{\text{BSFC}_{\max} - \text{BSFC}_{\min}} \tag{5.31}
$$

$$
d_{\text{CO}} = \frac{\text{CO}_{\max} - \text{CO}}{\text{CO}_{\max} - \text{CO}_{\min}} \tag{5.32}
$$

$$
d_{\text{UHC}} = \frac{\text{UHC}_{\max} - \text{UHC}}{\text{UHC}_{\max} - \text{UHC}_{\min}} \tag{5.33}
$$

$$d_{\text{NOx}} = \frac{\text{NOx}_{\text{max}} - \text{NOx}}{\text{NOx}_{\text{max}} - \text{NOx}_{\text{min}}} \tag{5.34}$$

The mathematical formulation used to transform multiple objective functions into a single-objective function for maximization is presented in Eqs. (5.30), (5.31), (5.32), (5.33), and (5.34). The present work involves the conflicting nature of output functions. The desirability function approach is applied to transform the conflicting nature of multiple outputs to a single output function aiming towards maximization. The fitness function value's composite desirability (D_0) is determined by considering all response functions using Eq. 5.29. The D_0 value varies between zero and 1. A D_0 value close to 1 signifies the ideal value for any process optimization.

5.6.1.3 Parametric Study of Optimization Algorithm

Step 4 The performance of six algorithms (GWO, PSO, TLBO, JAYA, DTBO, EBOA) was compared in terms of the convergence of the algorithms over successive iterations (refer to Fig. 5.3). All algorithms quickly reached close to the global fitness function value, which tends to have a steep curve at the beginning (refer to Fig. 5.4). The parametric study was conducted to tune the algorithm-specific parameters (GWO: a; PSO: inertia weight) and standard parameters (population size and maximum number of iterations) to maximize the fitness value. The study was conducted by varying individual parameters simultaneously while keeping the rest at the middle value of their respective levels. For example, a PSO parametric study determined the maximum fitness value by varying the inertia weight between their respective levels (0.1–0.9) after fixing the population size and maximum iteration to 50 and 250, respectively. The inertia weight corresponding to the maximized fitness value was recorded (0.5 for our study). Then, the population size was varied between 10 and 100 to locate the maximum fitness value while maintaining the fixed values of inertia weight and iteration at 0.5 and 250, respectively. Similarly, the maximum number of iterations was varied between 0 and 500 to locate the maximum fitness value after maintaining the optimized inertia weight and population size at 0.5 and 60, respectively. This procedure was repeated to conduct a parametric study of other algorithms.

The GWO and PSO algorithms required more than 100 iterations to converge (indicating a gradual descent) towards their highest personal fitness function values, or Do values, equal to 0.938 and 0.933, respectively. Although TLBO required fewer iterations to converge than PSO and GWO, it required more computation time due to its two-phase learning mechanism. The JAYA algorithm outperformed TLBO and PSO regarding computational time and convergence to the global fitness value. Note that the performances of the JAYA and GWO algorithms are comparable in terms of solution accuracy and computation time. Notably, the recently developed DTBO and EBOA algorithms showed identical results, and the highest fitness

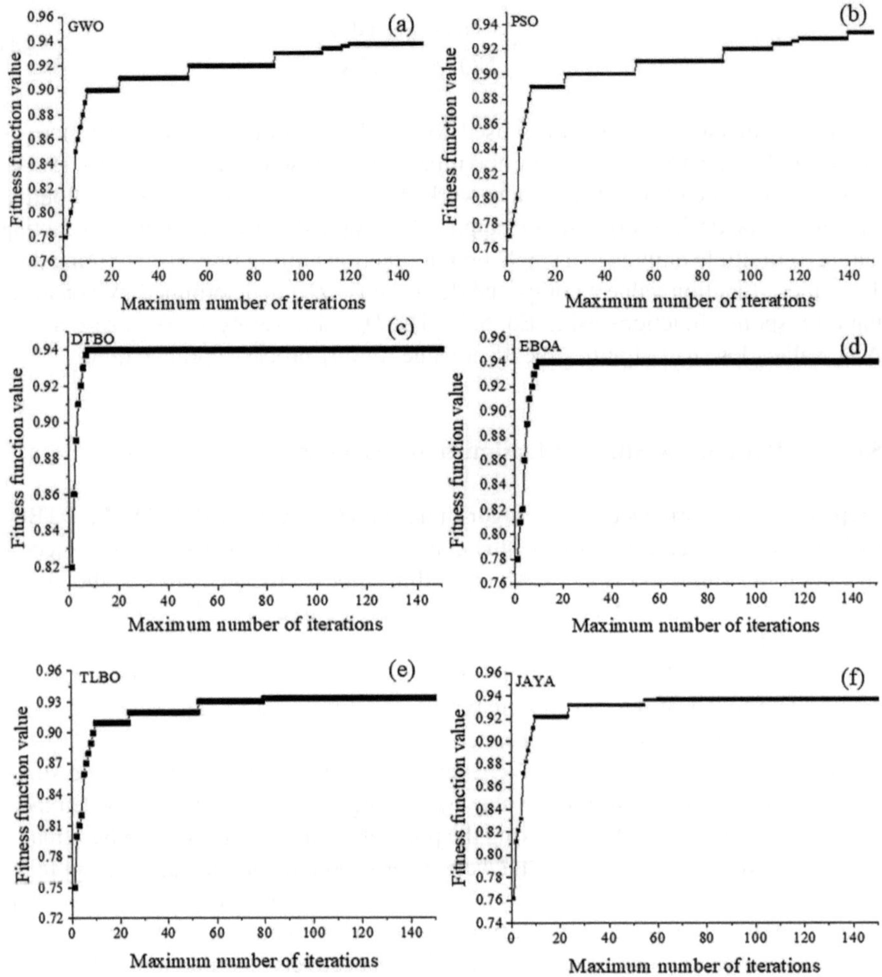

Fig. 5.3 Parametric study of optimization algorithms (example of Case 1): **(a)** GWO, **(b)** PSO, **(c)** DTBO, **(d)** EBOA, **(e)** TLBO, **(f)** JAYA

function values were equal to 0.94, with DTBO being more computationally efficient than EBOA. The computational time and fitness function value for each algorithm were found to be {20 s and 0.938} for GWO, {24 s and 0.933} for PSO, {26 s and 0.933} for TLBO, {20 s and 0.937} for JAYA, {6 s and 0.940} for DTBO, and {8 s and 0.940} for EBOA, respectively. The optimized algorithm parameters solved for case 1 as an example are presented in Table 5.4.

The computational efficiency of the algorithm varies due to algorithm complexity or learning mechanism (adaptive and dynamic) to locate the global fitness function, the requirement of tuning algorithm-specific parameters, and inappropriate initialization of algorithm parameters. DTBO and EBOA use adaptive and dynamic

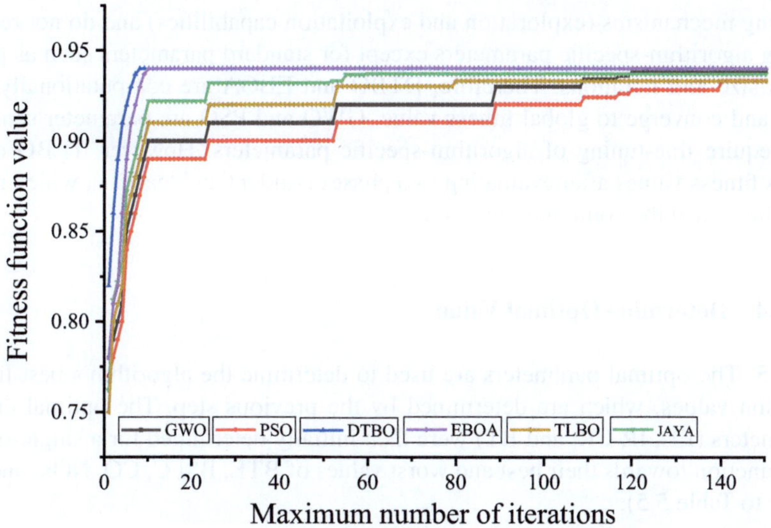

Fig. 5.4 Parametric study of all optimization algorithms

Table 5.4 Algorithm parameters and fitness function value

Algorithms	Tuning parameters (range)	Maximum fitness value
GWO	a: 1.2 PS: 50 Maximum iterations (0–500): 120 Computation time: 20 s	0.938
PSO	Inertia weight (0.1–0.9): 0.5 PS (10–100): 60 Maximum iterations (0–500): 140 Computation time: 24 s	0.933
TLBO	PS (10–100): 40 Maximum iterations (0–500): 80 Computation time: 26 s	0.933
JAYA	PS (10–100): 30 Maximum iterations (0–500): 60 Computation time: 20 s	0.937
DTBO	PS (10–100): 30 Maximum iterations (0–500): 8 Computation time: 6 s	0.940
EBOA	PS (10–100): 30 Maximum iterations (0–500): 10 Computation time: 7 s	0.940

learning mechanisms (exploration and exploitation capabilities) and do not require tuning algorithm-specific parameters except for standard parameters such as population size and iterations. Therefore, DTBO and EBOA are computationally efficient and converge to global fitness value. GWO and PSO are parameter sensitive and require fine-tuning of algorithm-specific parameters. However, TLBO determines fitness values after evaluating two phases (student and teacher), which might have increased the computational time.

5.6.1.4 Determine Optimal Value

Step 5 The optimal parameters are used to determine the algorithm's best fitness function values, which are determined by the previous step. The optimal engine parameters (EL, IP, CR, and BT) were then initially determined for a single-objective function towards their best and worst values of BTE, BSFC, CO, NOx, and HC (refer to Table 5.5).

Six case studies representing different weight fractions assigned to individual responses were used to determine the optimal conditions for conflicting requirements in multiple response functions. Each response's best and worst values (BTE,

Table 5.5 Summary results of optimal conditions for a single-objective function

	Best condition		Worst condition	
Optimization	Input variable	Output value	Input variable	Output value
BTE	EL: 100%	33.796%	EL: 20%	18.852%
	IP: 258.95 bar		IP: 160 bar	
	CR: 18		CR: 14	
	BT: 13.91 vol. %		BT: 30 vol. %	
BSFC	EL: 100%	0.210 kg/kWh	EL: 20%	0.493 kg/kWh
	IP: 235.28 bar		IP: 160 bar	
	CR: 18		CR: 14	
	BT: 20 vol. %		BT: 30 vol. %	
CO	EL: 20%	0.089 vol. %	EL: 94.19%	0.438 vol. %
	IP: 260 bar		IP: 180.69 bar	
	CR: 18		CR: 14	
	BT: 22.73 vol. %		BT: 10 vol. %	
UHC	EL: 20%	17.2 ppm	EL: 96.28%	70.2 ppm
	IP: 260 bar		IP: 160 bar	
	CR: 18		CR: 14	
	BT: 25.2 vol. %		BT: 10 vol. %	
NOx	EL: 20%	181.7 ppm	EL: 20%	1199.3 ppm
	IP: 160 bar		IP: 260 bar	
	CR: 14		CR: 18	
	BT: 10 vol. %		BT: 30 vol. %	

BSFC, CO, UHC, NOx) were used to determine the desirability value and later for composite desirability or fitness function value for all six case studies (refer Table 5.5). The results of six case studies are evaluated based on the composite desirability values to recommend single optimal conditions. Six algorithms determined the highest composite desirability values corresponding to each case. The differences in desirability values are attributed to the algorithmic learning and prediction mechanism. The highest desirability value determined for cases 1–6 equals 0.94, 0.969, 0.957, 0.944, 0.932, and 0.927, respectively. Case 2 is recommended as the optimal condition as their corresponding Do value is close to 1 (refer Table. 5.6).

5.6.1.5 Confirmation Experiments

Step 6 DTBO and EBOA determined identical optimal conditions that maximize the fitness or Do value. Case 2 resulted in the highest fitness function value and is recommended as an optimal condition that satisfies all outputs. Experiments were conducted for Case 2 based experimental conditions (EL: 100%, IP: 260 bar, CR: 18, and BT: 15.7 vol.%) resulting in BTE, BSFC, CO, NOx, and UHC equal to 32.5%, 0.259 kg/kWh, 0.212 vol.%, 514 ppm, and 38.1 ppm, respectively. The absolute percent deviation in prediction was found to be 3.91% for BTE, 1.93% for BSFC, 8.96% for CO, 3.86% for NOx, and 4.75% for UHC, respectively. The average absolute percent deviation in prediction is equal to 4.68%. Furthermore, the Ag_2O nanoparticles blended to biodiesel fuelled diesel engines correspond to optimal experimental conditions resulted in BTE, BSFC, CO, NOx, and UHC equal to 33.4%, 0.245 kg/kWhr, 0.202 Vol.%, 501 ppm, and 36.5 ppm, respectively. The absolute percent deviation in prediction with Ag_2O nanoparticles blended with biodiesel fuelled to diesel engines was found to increase by 2.77% for BTE, and decrease by 5.41% for BSFC, 4.72% for CO, 2.53% for NOx, and 4.20% for UHC, respectively. Therefore, the GGG-derived biodiesel can be deployed to run diesel engines with experimentally validated optimal conditions.

5.7 Concluding Remarks

The transesterification process transforms the Garcinia Gummi-Gutta seed oil into biodiesel conversion. Furthermore, the biodiesel fuelled diesel engines were evaluated for performance and emission characteristics. The experiments were planned to detail the process insights and derive empirical equations for process optimization. Multi-objective optimization tasks were conducted to determine the optimal conditions for maximizing BTE and minimizing BSFC, CO, NOx, and UHC emissions. Six algorithms (GWO, PSO, DTBO, EBOA, TLBO, and JAYA) were applied to perform the said tasks. Six case studies were conducted to recommend optimal conditions that satisfy all responses. The following conclusions were drawn:

Table 5.6 Case studies for identifying optimal conditions for multi-objective optimization

		Input variables				Output variables					Fitness function or desirability value
Case study	Algorithms	EL (%)	IP (Bar)	CR	BT (Vol. %)	BTE (%)	BSFC, kg/kWh	CO (Vol. %)	UHC (ppm)	NOx (ppm)	
Case 1	TLBO	95	242	18	18	32.67	0.287	0.186	36.88	605.54	0.933
	GWO	100	260	17.8	16.5	33.57	0.271	0.192	39.08	560.04	0.938
	PSO	95	242	18	18	32.67	0.287	0.186	36.88	605.54	0.933
	DTBO	100	260	18	16.4	33.74	0.270	0.185	38.71	557.60	0.940
	EBOA	100	260	18	16.3	33.74	0.269	0.186	38.87	554.15	0.940
	JAYA	98	246	18	16.2	33.35	0.269	0.199	39.50	546.36	0.937
Case 2	TLBO	100	221	18	16	33.66	0.265	0.212	40.60	534.95	0.966
	GWO	100	213.4	18	16.8	33.58	0.272	0.205	39.51	558.54	0.966
	PSO	100	230	18	15.9	33.70	0.263	0.210	40.49	534.87	0.967
	DTBO	100	260	18	15.7	33.77	0.264	0.193	39.91	533.86	**0.969**
	EBOA	100	260	18	15.7	33.77	0.264	0.193	39.91	533.86	**0.969**
	JAYA	100	245	18	15.9	33.74	0.264	0.202	40.02	538.67	0.968
Case 3	TLBO	100	221	18	10	33.59	0.211	0.308	54.29	370.46	0.953
	GWO	100	235	18	10	33.63	0.210	0.301	53.82	373.37	0.954
	PSO	100	221	18	10	33.59	0.211	0.308	54.29	370.46	0.953
	DTBO	100	260	18	10	33.65	0.212	0.282	52.98	374.39	0.957
	EBOA	100	260	18	10	33.65	0.212	0.282	52.98	374.39	0.957
	JAYA	100	250	18	10	33.65	0.211	0.291	53.32	374.63	0.956
Case 4	TLBO	20	230	18	21	24.43	0.385	0.111	19.93	655.34	0.941
	GWO	20	230	18	21	24.43	0.385	0.111	19.93	655.34	0.942
	PSO	20	230	18	21	24.43	0.385	0.111	19.93	655.34	0.941
	DTBO	20	243	18	20.7	24.52	0.383	0.105	19.68	651.39	0.944
	EBOA	20	243	18	20.7	24.52	0.383	0.105	19.68	651.39	0.944
	JAYA	20	232	18	20.8	24.47	0.384	0.110	20.01	648.32	0.942

Case 5	TLBO	20	230	18	25	23.70	0.406	0.110	18.29	840.10	0.932
	GWO	20	230	18	25	23.70	0.406	0.110	18.29	840.10	0.932
	PSO	20	232	18	25.2	23.67	0.407	0.110	18.22	851.74	0.931
	DTBO	20	235	18	25	23.73	0.406	0.108	18.11	844.08	0.932
	EBOA	20	235	18	25	23.73	0.406	0.108	18.11	844.08	0.932
	JAYA	20	233	18	25	23.72	0.406	0.109	18.19	842.51	0.932
Case 6	TLBO	20	249	14	10	24.12	0.376	0.281	52.29	191.89	0.926
	GWO	20	250	14	10	24.12	0.376	0.280	52.25	191.62	0.926
	PSO	20	251	14	10	24.12	0.376	0.279	52.22	191.34	0.926
	DTBO	20	254	14	10	24.12	0.376	0.275	52.10	190.45	0.927
	EBOA	20	254	14	10	24.12	0.376	0.275	52.10	190.45	0.927
	JAYA	20	251	14	10	24.12	0.376	0.279	52.22	191.34	0.926

1. All algorithms (GWO, PSO, DTBO, EBOA, TLBO, and JAYA) determined an identical optimal condition that maximizes the biodiesel yield. The optimal conditions (methanol: 40 mL, reaction time: 75 min, sodium hydroxide concentration: 0.83 g) resulted in a maximum biodiesel conversion of 94.01%. The resulting experimental biodiesel yield was found to be equal to 96.2%.
2. CCD experiments were conducted, and an empirical equation represented outputs (BSFC, BTE, CO, NOx, and UHC) as a function of EL, IP, CR, and BT. The empirical equations resulted in a good coefficient of determination with a value equal to 0.9940 for BTE, 0.9937 for BSFC, 0.9745 for CO, 0.9863 for UHC, and 0.985 for NOx, respectively. Therefore, all models are efficient to apply for prediction and optimization.
3. For single and multiple objective optimizations, six algorithms (GWO, PSO, TLBO, JAYA, DTBO, and EBOA) were applied. The global best and worst values of individual outputs were determined for each response and found to equal 33.796% and 18.85% for BTE, 0.21 kg/kWh and 0.493 kg/kWh for BSFC, 0.089 vol.% and 0.438 vol.% for CO, 17.2 ppm and 70.2 ppm for UHC, and 181.7 ppm and 1199.3 ppm, respectively.
4. Multi-objective optimizations have many potential solutions, relying on the weight fractions assigned to individual responses. Six cases (equal weight fraction for six responses (1/5 = 0.2) correspond to case 1, cases 2–6 assigned with maximum weight fraction (0.6) for individual output with least for the rest (0.1) towards BTE, BSFC, CO, UHC, and NOx. All six cases were evaluated based on composite desirability value to recommend the single optimal conditions that satisfy all output.
5. A parametric study for each algorithm was conducted, taking case 1 as an example. The computational time and fitness function value for each algorithm were found to be equal to {20 s and 0.938} for GWO, {24 s and 0.933} for PSO, {26 s and 0.933} for TLBO, {20 s and 0.937} for JAYA, {6 s and 0.940} for DTBO, and {8 s and 0.940} for EBOA, respectively. The DTBO and EBOA converge to a global fitness function value equal to 0.94, with DTBO being more computationally efficient than EBOA.
6. DTBO and EBOA determined conditions correspond to case 2 and are recommended as optimal ones that satisfy all outputs. Experimental condition (EL: 100%, IP: 260 bar, CR: 18, and BT: 15.7 vol.%) resulted in BTE, BSFC, CO, NOx, and UHC equal to 32.5%, 0.259 kg/kWh, 0.212 vol.%, 514 ppm, and 38.1 ppm, respectively. The average absolute percent deviation in prediction is equal to 4.68%.
7. Ag_2O nanoparticles blended with to biodiesel fuelled diesel engines were found to increase by 2.77% for BTE and decrease by 5.41% for BSFC, 4.72% for CO, 2.53% for NOx, and 4.20% for UHC, respectively.
8. DTBO and EBOA recommended optimal conditions that can be used to run diesel engines for better emission and performance characteristics.

Scope for Future Research Directions

The research study can be extended in three aspects:

- To ensure a consistent and sustainable supply of feedstock for biodiesel production, future research can focus on optimizing the blending ratios of GGG oil with other locally available oils.
- The application of advanced multi-layered coatings on critical engine components, such as pistons, cylinders, and valves, can be explored to enhance engine durability, reduce frictional losses, and improve thermal efficiency.
- The authors can use other parameters (injection timing, injection pressure, injection angle, exhaust gas recirculation, air-to-fuel ratio, stroke length, intake air temperature, coolant temperature, throttle position, and cylinder bore diameter) for optimizing the engine parameters for better thermal performance and emission characteristics.

References

1. Manjunath Patel G.C., et al.: Optimization of machining of hard material. pp. 103–128 (2020)
2. Sarker, I.H.: Machine learning: algorithms, real-world applications and research directions. SN Comput. Sci. **2**(3), 160 (2021)
3. Mirjalili, S., Mirjalili, S.M., Lewis, A.: Grey wolf optimizer. Adv. Eng. Softw. **69**, 46–61 (2014)
4. Mirjalili, S., et al.: Multi-objective grey wolf optimizer: a novel algorithm for multi-criterion optimization. Exp. Syst. Appl. **47**, 106–119 (2016)
5. Kennedy, J., Eberhart, R.: Particle swarm optimization. In: Proceedings of ICNN'95-international conference on neural networks. IEEE (1995)
6. Ding, Y., et al.: The accuracy and efficiency of GA and PSO optimization schemes on estimating reaction kinetic parameters of biomass pyrolysis. Energy. **176**, 582–588 (2019)
7. Rao, R.V., Savsani, V.J., Vakharia, D.: Teaching–learning-based optimization: a novel method for constrained mechanical design optimization problems. Comput. Aided Des. **43**(3), 303–315 (2011)
8. Rao, R.V., Savsani, V.J., Vakharia, D.: Teaching–learning-based optimization: an optimization method for continuous non-linear large scale problems. Inform. Sci. **183**(1), 1–15 (2012)
9. Rao, R.V., Waghmare, G.G.: A new optimization algorithm for solving complex constrained design optimization problems. Eng. Optim. **49**(1), 60–83 (2017)
10. Manjunath Patel G.C., Chate, G.R., Parappagoudar, M.B: Modelling and optimization of alpha-set sand moulding system using statistical design of experiments and evolutionary algorithms. pp. 1–28 (2020)
11. Patel, G.M., Jagadish: Experimental modeling and optimization of surface quality and thrust forces in drilling of high-strength Al 7075 alloy: CRITIC and meta-heuristic algorithms. J. Braz. Soc. Mech. Sci. Eng. **43**, 1–21 (2021)
12. Rao, R.V., Savsani, V., Balic, J.: Teaching–learning-based optimization algorithm for unconstrained and constrained real-parameter optimization problems. Eng. Optim. **44**(12), 1447–1462 (2012)
13. Sibalija, T.V., et al.: A soft computing-based study on WEDM optimization in processing Inconel 625. Neural Comput. Appl. **33**(18), 11985–12006 (2021)
14. Patel, G.M., et al.: Teaching learning based optimization of squeeze casting process for quality castings. In: IOP Conference Series: Materials Science and Engineering. IOP Publishing (2018)

15. Ajith, B., et al.: Analysis and optimisation of transesterification parameters for high-yield Garcinia Gummi-Gutta biodiesel using RSM and TLBO. Aust. J. Mech. Eng. **20**(5), 1463–1478 (2022)
16. Jagadish, et al.: Abrasive water jet machining for a high-quality green composite: the soft computing strategy for modeling and optimization. J. Braz. Soc. Mech. Sci. Eng. **44**(3), 83 (2022)
17. Rao, R.V., Kalyankar, V.D.: Parameter optimization of modern machining processes using teaching–learning-based optimization algorithm. Eng. Appl. Artific. Intellig. **26**(1), 524–531 (2013)
18. Rao, R.: Jaya: a simple and new optimization algorithm for solving constrained and unconstrained optimization problems. Int. J. Indus. Eng. Comput. **7**(1), 19–34 (2016)
19. Rao, R.V.: Jaya: an advanced optimization algorithm and its engineering applications (2019)
20. Trojovský, P., Dehghani, M.J.P.C.S.: A new optimization algorithm based on mimicking the voting process for leader selection. PeerJ Comput Sci. **8**, e976 (2022)
21. Dharmegowda, I.Y., et al.: Optimization for waste coconut and fish oil derived biodiesel with MgO nanoparticle blend: Grey relational analysis, grey wolf optimization, driving training based optimization and election based optimization algorithm. Fuel. **338**, 127249 (2023)
22. Dehghani, M., Trojovská, E., Trojovský, P.: A new human-based metaheuristic algorithm for solving optimization problems on the base of simulation of driving training process. Sci. Rep. **12**(1), 9924 (2022)
23. Rehman, H., et al.: Driving training-based optimization (dtbo) for global maximum power point tracking for a photovoltaic system under partial shading condition. IET Renew. Power Gener. **17**(10), 2542–2562 (2023)
24. Patel, G.M., et al.: Multi-objective optimization of squeeze casting process using genetic algorithm and particle swarm optimization. Arch. Found. Eng. **16** (2016)
25. Lü, X., et al.: Energy management of hybrid electric vehicles: a review of energy optimization of fuel cell hybrid power system based on genetic algorithm. Energy Convers. Manag. **205**, 112474 (2020)
26. Chate, G.R., et al.: Comprehensive modelling, analysis and optimization of furan resin-based moulding sand system with sawdust as an additive. J. Braz. Soc. Mech. Sci. Eng. **41**, 1–24 (2019)
27. Shettigar, A.K., et al.: Artificial bee colony, genetic, back propagation and recurrent neural networks for developing intelligent system of turning process. SN Appl. Sci. **2**, 1–21 (2020)
28. Stützle, T., et al.: Parameter adaptation in ant colony optimization. In: Autonomous Search, pp. 191–215 (2012)
29. Alsewari, A.A., et al.: Combinatorial test list generation based on Harmony Search Algorithm. J. Ambient Intell. Hum. Comput., 1–17 (2020)
30. Rezaee Jordehi, A.: A chaotic-based big bang–big crunch algorithm for solving global optimisation problems. Neural Comput. Appl. **25**, 1329–1335 (2014)
31. Belloufi, A., et al.: Intelligent selection of machining parameters in multipass turnings using firefly algorithm. Modell. Simul. Eng. **2014**, 8–8 (2014)
32. Khalilpourazari, S., Khalilpourazary, S.J.: Optimization of production time in the multi-pass milling process via a Robust Grey Wolf Optimizer. Neural Comput. Appl. **29**, 1321–1336 (2018)
33. Venkataramana, S.H., et al.: Niger seed oil-based biodiesel production using transesterification process: experimental investigation and optimization for higher biodiesel yield using box–behnken design and artificial intelligence tools. Appl. Sci. **12**(12), 5987 (2022)

MIX
Papier aus verantwortungsvollen Quellen
Paper from responsible sources
FSC® C105338

If you have any concerns about our products,
you can contact us on
ProductSafety@springernature.com

In case Publisher is established outside the EU,
the EU authorized representative is:
Springer Nature Customer Service Center GmbH
Europaplatz 3, 69115 Heidelberg, Germany

Printed by Libri Plureos GmbH
in Hamburg, Germany